Mastering
Chemistry

Palgrave Master Series

Accounting
Accounting Skills
Advanced English Language
Advanced Pure Mathematics
Arabic
Basic Management
Biology
British Politics
Business Communication
Business Environment
C Programming
C++ Programming
Chemistry
COBOL Programming
Communication
Computing
Counselling Skills
Counselling Theory
Customer Relations
Database Design
Delphi Programming
Desktop Publishing
e-Business
Economic and Social History
Economics
Electrical Engineering
Electronics
Employee Development
English Grammar
English Language
English Literature
Fashion Buying and Merchandising
  Management
Fashion Marketing
Fashion Styling
Financial Management
Geography
Global Information Systems

Globalization of Business
Human Resource Management
Information Technology
International Trade
Internet
Java
Language of Literature
Management Skills
Marketing Management
Mathematics
Microsoft Office
Microsoft Windows, Novell
  NetWare and UNIX
Modern British History
Modern European History
Modern United States History
Modern World History
Networks
Novels of Jane Austen
Organisational Behaviour
Pascal and Delphi Programming
Philosophy
Physics
Poetry
Practical Criticism
Psychology
Public Relations
Shakespeare
Social Welfare
Sociology
Spanish
Statistics
Strategic Management
Systems Analysis and Design
Team Leadership
Theology
Twentieth-Century Russian History
Visual Basic
World Religions

www.palgravemasterseries.com

Palgrave Master Series
Series Standing Order ISBN 0–333–69343–4
(outside North America only)

You can receive future titles in this series as they are published by placing a standing order. Please contact your bookseller or, in case of difficulty, write to us at the address below with your name and address, the title of the series and the ISBN quoted above.

Customer Services Department, Macmillan Distribution Ltd
Houndmills, Basingstoke, Hampshire RG21 6XS, England

Mastering
Chemistry

Second Edition

Peter D. Riley

MACMILLAN

First published 1982 as *Mastering Chemistry* by Peter Critchlow
Reprinted 16 times

This edition published 2000 by
MACMILLAN PRESS LTD
Houndmills, Basingstoke, Hampshire RG21 6XS
and London
Companies and representatives
throughout the world

ISBN-10: 0-333-69598-4
ISBN-13: 978-0-333-69598-2

A catalogue record for this book is available
from the British Library.

This book is printed on paper suitable for recycling and made from fully managed and sustained forest sources.

10   9   8   7   6   5   4
09 08 07 06

Printed in China

To my granddaughter Megan Kate

Contents

◼ ⌄ Preface

This second edition of *Mastering Chemistry* has been completely rewritten to meet the syllabus changes which have taken place since the first edition was published in 1982.

In this book I have tried to present the topics for chemistry at introductory level in a progression that begins with simple concepts and ideas such as the nature of matter and its organisation in the structure of the Earth and its atmosphere, then moving onto more complex concepts such as atomic structure and the nature of chemical reactions, their uses in industry and implications for the environment. By presenting the topics in this way I hope that readers with an interest in chemistry but little formal tuition in the subject may build up their knowledge and that students may find the book helpful in their preparation for examination.

Each chapter begins with a list of objectives to help the reader focus on the topic. There are questions throughout the text which help in the development of a range of comprehension skills. Questions involving calculations have answers at the back of the book (p. 254). Each chapter ends with a summary in which each point is page referenced to help with revision. I hope the arrangement of the chapters and the features they contain for active learning help you to master chemistry.

PETER RILEY

Acknowledgements

Peter Riley and the publishers would like to thank the following for providing photographs for this book: Derek Whitford; Peter Hartley; Merck Ltd; F8 Imaging and the British Geological Survey.

Peter Riley would also like to thank G. M. Cartledge for the constructive criticism on the early draft of the manuscript.

⬛ V̌ ▮ The states of matter

Objectives

When you have completed this chapter you should be able to:
- Describe the properties of the **three states of matter**
- Describe how **matter changes states**
- Understand the **Gas Laws** and be able to use them in calculations
- Understand the **kinetic theory** and use it to explain the states of matter and how they change
- Explain **diffusion, dissolving and density** in terms of the kinetic theory.

1.1 Where matter came from

About 15 billion years ago an explosion took place called the 'Big Bang'. This explosion formed the universe. In the first tiny fraction of a second after the explosion, atoms of hydrogen and helium formed and spread out in all directions. Many of them were eventually pulled together by the force of gravity to form stars. The atoms of other elements formed in the stars and were released into space when the stars no longer released much energy as light and heat or when the star exploded. In time, hydrogen, helium and dust made by other elements gathered in a cloud by the force of gravity and formed the Solar system. The Sun contained most of the hydrogen and helium while the other elements formed the planets, comets and asteroids. Living things are formed from many of the elements present on the Earth. The chemicals taking part in reactions in your body at this moment formed in stars millions of years ago.

1.2 The three states of matter

Atoms and molecules can exist in three types of forms. These forms are called the three states of matter – solid, liquid and gas. They can be seen in Figure 1.1.

In chemical equations it is conventional to use these subscripts for the states of matter after each chemical formula. The subscripts are solid (s), liquid (l) and gas (g).

Figure 1.1   A sea scape

The physical properties of each state differ from the physical properties of the other two states, as the following descriptions show.

(i) Solids

Solids – for example, rocks and metals – have a fixed shape and occupy a certain volume of space. They cannot be squashed to take a smaller volume but may have pressure applied to them to change their shape. A solid maintains its properties over a range of temperatures. Within this range a solid will usually expand a very small amount if its temperature is increased and will contract a very small amount if its temperature is reduced. At a certain temperature a pure solid loses its physical properties and becomes a liquid. This temperature is known as the **melting point** of the solid. The ice in Figure 1.2 is about to melt.

(ii) Liquids

Liquids – for example, water and oil – occupy a certain volume of space but do not have a fixed shape. A liquid can be poured and takes up the shape of its container. It can also be made to flow if pressure is applied to it but cannot be squashed to take a smaller volume. A liquid maintains its properties over a range of temperatures. Within this range a liquid will usually expand a small amount if

Figure 1.2 Ice

the temperature is increased and will contract a small amount if the temperature is reduced. At a certain temperature a pure liquid loses its physical properties and becomes a gas. The temperature at which this occurs is known as the **boiling point** of the liquid.

(iii) Gases

Gases – for example, carbon dioxide and methane – do not have a definite volume or a fixed shape. A gas can be made to flow with a small amount of pressure and can be poured. It fills any container into which it is placed and can be compressed greatly to take up a very small volume of space. If the temperature of a gas is raised it expands a large amount and if the temperature of the gas is lowered it also contracts a large amount.

QUESTIONS

1 What happens to a substance when its temperature rises (a) above its melting point, (b) above its boiling point?
2 How does the volume of a solid differ from the volume of a gas?
3 How does the shape of a solid differ from the shape of a liquid?
4 Construct a table in which the physical properties of the three states of matter are compared.

1.3 Heating and cooling curves

(i) A heating curve

If a solid is heated and observed while its temperature is being taken a graph can be produced as shown in Figure 1.3. This is called a **heating curve**.

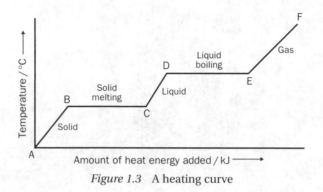

Figure 1.3 A heating curve

Line A–B shows the temperature change as the solid is heated up.
Line B–C shows the temperature as the solid melts. The temperature does not rise at this time even though the material is receiving the same amount of heat as in the earlier part of the experiment. The heat is being absorbed by the material and used to bring about the change of state from solid to liquid. This process in which energy is absorbed is called an **endothermic process**.
Line C–D shows the temperature change as the liquid heats up.
Line D–E shows the temperature as the liquid boils.

(ii) A cooling curve

If a boiling liquid is cooled down and observed while its temperature is being taken a graph can be produced, as shown in Figure 1.4. It is called a **cooling curve**.

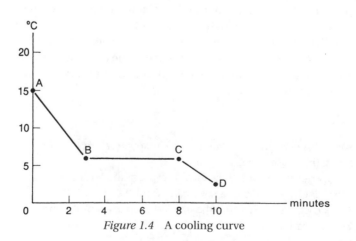

Figure 1.4 A cooling curve

Line A–B shows the temperature of the liquid as it cools.
Line B–C shows the temperature of the liquid as it freezes and turns into a solid.
This temperature is called the **freezing point** and is the same as the melting point.
The temperature does not fall as the material turns into a solid even though the
material is not receiving heat. This process in which the material is giving out
heat is called an **exothermic process**.
Line C–D shows the temperature change as the solid cools.

QUESTION

5 List the order in which the changes of state occur when (a) a heating curve, (b)
a cooling curve is being plotted.

1.4 Other ways in which a state may change

(i) Evaporation

When a liquid is left exposed to the air its volume reduces in time. This is due to
the process of **evaporation**. In this process some of the liquid at the liquid surface
turns into a gas and mixes with the other gases in the air. The rate of evaporation
increases as the temperature of the liquid increases until the liquid reaches
boiling point.

(ii) Condensation

A gas may be changed into a liquid by lowering its temperature. This process of
changing from gas to liquid is called **condensation**. A common example of condensation is the formation of water droplets on a cold window pane. This occurs
when a person breathes on the glass. Water vapour (a gas) in the warm breath
condenses to form the drops of water.

(iii) Sublimation

Sublimation is a relatively rare process. It occurs when a solid turns directly into
a gas without going through the liquid state. Solid carbon dioxide (also known as
'dry ice') sublimes to form carbon dioxide gas. It is used to create mists in theatre
productions and rock concerts.

Sublimation also occurs when a gas turns directly into a solid. Hot sulphur
leaving the vent of a volcano is in the form of a gas. It sublimes to form a crust
on the rocks around the vent.

1.5 Changes in the states of matter

The way the states of matter change can be summarised as shown in
Figure 1.5.

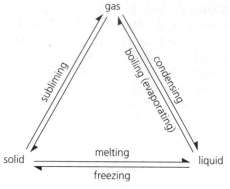

Figure 1.5 States of matter triangle

1.6 Diffusion

When food is being cooked some of the substances form gases which mix with the air. You can detect these gases as the smells of cooking. They have mixed with the air and travelled from the food to your nose. This process of mixing and moving is called **diffusion**.

In the laboratory the diffusion of gases is usually demonstrated by setting up a gas jar of bromine and a gas jar of air as Figure 1.6 shows.

In about 24 hours the bromine and air have completely mixed due to diffusion.

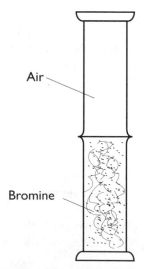

Figure 1.6 Gas jar of bromine below a gas jar of air

(ii) Comparing rates of diffusion

Ammonia gas evaporates from concentrated ammonia solution and hydrogen chloride gas evaporates from concentrated hydrochloric acid. When the two

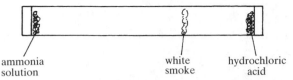

Figure 1.7 Gas tube with cotton wool pads at each end and cloud nearer acid end

gases meet they form ammonium chloride – a white solid which makes a cloud in the air.

When the apparatus is set up as shown in Figure 1.7 the gases escape from the cotton wool at each end of the tube and the white cloud forms closer to the cotton wool soaked in concentrated hydrochloric acid. This experiment shows that ammonia diffuses through air more quickly than hydrogen chloride.

QUESTION

6 Where would the cloud have formed if (a) both gases diffused at the same rate and (b) hydrogen chloride diffused faster than ammonia?

1.7 Boyle's Law and Charles' Law

In the seventeenth and eighteenth century due to the work of Robert Boyle (1627–91) and Jacques Charles (1746–1823) it was realised that the condition of any gas could be altered if a change was made to its temperature, volume and pressure.

(i) Boyle's Law

Robert Boyle discovered that as the pressure on a gas was increased its volume was reduced in the following way if the temperature was kept constant. If the gas pressure was doubled the volume was reduced by half of its original volume. If the gas pressure was trebled the volume was reduced to a third of the original volume. This can be expressed as the equation

$P \times V = Constant$

and is known as **Boyle's Law**

It can also be expressed as

$P_1 \times V_1 = P_2 \times V_2$

where P_1 and V_1 are the first pressure and volume of the gas and P_2 and V_2 are the second pressure and volume of the gas.

Boyle's Law can be used to calculate an unknown pressure or volume, as the following example shows.

Pressure can be measured in atmospheres or pascals (Pa)

(*Note:* 1 atmosphere $= 1 \times 10^5$ Pa)

EXAMPLE

200 cm^3 of gas at constant temperature at 1 atmosphere has its pressure increased to 3 atmospheres. What is its new volume?

Rearranging the equation to

$$V_2 = \frac{P_1 V_1}{P_2} = \frac{1 \times 200}{3} = 66.7\,cm^3$$

QUESTIONS

7 A gas with a volume of 500 cm^3 at constant temperature at 1 atmosphere has its pressure increased to 5 atmospheres. What is its new volume?

8 A gas has a volume of 600 cm^3 at constant temperature and a pressure of 1 atmosphere. What is its pressure when its volume is reduced to 200 cm^3?

(ii) Charles' Law

Jacques Charles measured the volume of gases very accurately as he heated and cooled them. He kept the pressure constant in these experiments and found that as the temperature of the gas increased its volume also increased. This can be expressed as the equation

$$\frac{V}{T} = \text{constant} \quad \text{or} \quad \frac{V_1}{T_1} = \frac{V_2}{T_2}$$

where V_1 and T_1 are the first volume and temperature of the gas and V_2 and T_2 are the second volume and temperature of the gas.

This relationship between the volume and temperature of a gas at constant pressure is called **Charles' Law**.

Charles also discovered that when the temperature of a gas was raised or lowered by 1°C its volume increased or decreased by 1/273rd of the volume he recorded for the gas at 0°C. He believed this meant that if he kept cooling the gas below 0°C it would eventually shrink to zero volume at –273°C, Lord Kelvin (1824–1907) studied the work of Charles and suggested that the volume did not shrink to zero but that the atoms or molecules in the gas stopped moving (see p. 14).

Kelvin suggested that a scale of temperature be set up with –273°C being equal to zero on the new scale. This scale is known as the **Kelvin scale**. On this scale ice melts at 273 K and water boils at 373 K. The Kelvin scale is used in calculations with the gas laws.

QUESTIONS

9 Convert these temperatures on the Celsius scale to temperatures on the Kelvin scale (a) 20°C, (b) 35°C, (c) 250°C, (d) –10°C, (e) –57°C.

10 Convert these temperatures on the Kelvin scale to temperatures on the Celsius scale (a) 300 K, (b) 350 K, (c) 500 K, (d) 200 K, (e) 10 K.

1.8 Calculations with Charles' Law

Charles' Law can be used to calculate an unknown volume or temperature, as the following examples show.

EXAMPLES

1 One litre of gas at a temperature of 283 K is heated to 383 K.
   What is its new volume if its pressure remains constant?
   The equation

   $$\frac{V_1}{T_1} = \frac{V_2}{T_2}$$

   can be rearranged to

   $$V_2 = \frac{V_1 \times T_2}{T_1}$$

   $$V_2 = 1 \times \frac{373}{273} = 1.4 \text{ litres}$$

2 A gas of volume 50 cm^3 at 273 K rises to a volume of 150 cm^3 at constant pressure. What temperature is the increased volume of gas?
   The equation

   $$\frac{V_1}{T_1} = \frac{V_2}{T_2}$$

   can be rearranged to

   $$T_2 = \frac{V_2 \times T_1}{V_1}$$

   $$T_2 = \frac{150}{50} \times 273 = 819 \text{ K}$$

QUESTIONS

11 2 litres of a gas at a temperature of 290 K is heated to 340 K. What is its new volume if its pressure remains constant?

12 A gas of volume 2 litres has a temperature of 273 K. Its volume rises to 3 litres while its pressure is kept constant. What temperature is the increased volume of gas?

13 A volume of gas at 280 K has a temperature rise to 360 K. At this new temperature its volume is 2.5 litres. What was the original volume of the gas?

1.9 The Gas Law

Charles' Law and Boyle's Law are combined in the general **Gas Law**.
   This can be expressed as the equation:

$$P \times V = \text{Constant or as } \frac{P_1 \times V_1}{T_1} = \frac{P_2 \times V_2}{T_2}$$

where P = pressure of the gas, V = volume of the gas and T = temperature of the gas on the Kelvin scale.

The Gas Law can be used to calculate the unknown pressure, volume or temperature of a gas, as the following examples show.

EXAMPLES

1 A 2 litre volume of gas with a pressure of 1 atmosphere and a temperature of 400 K has its volume kept constant but its temperature raised to 500 K. What is the new pressure of the gas.

The equation is re arranged to

$$P_2 = \frac{P_1 \times V_1 \times T_2}{T_1 \times V_2}$$

When the numbers in the question are placed in the equation, the new pressure can be calculated as follows:

$$P_2 = \frac{1 \times 2 \times 500}{400 \times 2} = 1.25 \text{ at mospheres}$$

2 A 3 litre volume of gas at 200 K has a pressure of 5 atmospheres. Its pressure is kept constant but its temperature is raised to 700 K. What is the new volume of the gas?

The equation is rearranged to

$$V_2 = \frac{P_1 \times V_1 \times T_2}{T_1 \times P_2}$$

When the numbers in the question are placed in the equation the new pressure can be calculated as follows:

$$V_2 = \frac{5 \times 3 \times 700}{200 \times 5} = 10.5 \text{ l}$$

3 A 6 litre volume of gas at 450 K has a pressure of 4 atmospheres. Its pressure is reduced to 1 atmosphere but the volume is kept the same. What is the new temperature of the gas?

The equation is rearranged to

$$T_2 = \frac{P_2 \times V_2 \times T_1}{P_1 \times V_1}$$

When the numbers in the question are placed in the equation, the new pressure can be calculated as follows:

$$T_2 = \frac{1 \times 6 \times 450}{4 \times 6} = 112.5 \text{ K}$$

14 A 25 litre volume of gas with a pressure of 4 atmospheres and a temperature of 300 K has its volume kept constant but its temperature raised to 500 K. What is the new pressure of the gas?

15 A 6 litre volume of gas at 150 K has a pressure of 3 atmospheres. Its pressure is kept constant but its temperature is raised to 300 K. What is the new volume of the gas?

16 A 7 litre volume of gas at 800 K has a pressure of 2 atmospheres. Its pressure is kept constant but its temperature is lowered to 300 K. What is the new volume of the gas?

17 A 2 litre volume of gas at 200 K has a pressure of 10 atmospheres. Its pressure is reduced to 6 atmospheres but the volume is kept the same. What is the new temperature of the gas?

18 A 3 litre volume of gas at 360 K has a pressure of 5 atmospheres. Its pressure is increased to 7 atmospheres and the volume is reduced to 2 litres. What is the new temperature of the gas?

1.10 Matter and particles

Robert Brown (1773–1858) used a microscope to study pollen grains. When they were mounted in water he saw the grains move about jerkily and thought it was due to some form of life in them. When Brown tested the water with particles of dye from a non-living substance he found they moved too and concluded that something in the water made the tiny objects move. This movement became known as **Brownian motion**.

James Clerk Maxwell (1813–79) and Ludwig Boltzmann (1844–1906) studied gases and found a way to describe how gases behave by considering them to be made of tiny particles. They developed their ideas into the **Maxwell–Boltzmann kinetic theory of gases**. Later this theory was expanded to include liquids and solids and today is usually known simply as the **kinetic theory**.

The kinetic theory states that:

- matter is made from tiny particles
- the particles at a temperature above −273 K are moving all the time. They possess the energy of motion which is called **kinetic energy**
- the rate of movement of the particles is linked to their temperature, the particles have low kinetic energy and move slowly at a low temperature and have higher kinetic energy and move more quickly at a higher temperature
- particles can be attracted to each other
- particles can be different weights and sizes
- heavier particles move more slowly than more lightweight particles.

The kinetic theory can be used to explain the properties of matter described on pp. 2 and 3. It may be useful to look back at these pages as you read how the kinetic theory relates to them.

1.11 Explanations with the kinetic theory

A THE SOLID STATE

In solids the particles are strongly attracted to each other. They become packed together. The way the particles pack together in many solids give the solid a particular flat sided or crystalline shape (see Figure 1.8).

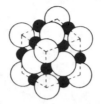

Figure 1.8 Particles packed together as in a salt crystal

Although each particle has a fixed position in the solid it still moves by vibrating about its centre, as Figure 1.9a shows.

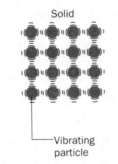

Figure 1.9a Particles vibrating in a solid

(i) Expansion

When a solid is heated the particles acquire more kinetic energy. This makes them move further about their fixed point and with greater speed. As the particles move a little further this makes the solid expand or increase in volume.

(ii) Melting

If the solid is heated strongly enough the particles acquire so much kinetic energy that they break away from their fixed points and slide around each other. When this happens the solid melts and becomes a liquid.

B THE LIQUID STATE

In liquids the particles have more kinetic energy than particles in solids. This means that they are not held in a fixed position and can slide around each another, as shown in Figure 1.10.

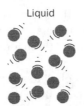

Liquid

Figure 1.10 Particles moving in a liquid

QUESTION

19 Look at the heating and cooling curves in Figure 1.3 and 1.4 and describe the changes that take place in a group of particles when the solid is heated and a liquid is cooled.

(i) Brownian movement

It is the movement of the particles in a liquid that produces Brownian movement. The particles push against any small solids such as pollen grains and make them move about in the liquid.

(ii) Freezing

If a liquid is cooled its particles lose kinetic energy. This makes them move more slowly. If the liquid is cooled down enough each particle stops moving and vibrates about a fixed point. When this happens the liquid has frozen and turned into a solid.

(iii) Dissolving

A substance which dissolves in a liquid is called a **solute**. The liquid in which the solute dissolves is called a **solvent**. When a solute dissolves in a solvent it does so because the solute's particles can break up and fill spaces between the particles of the the solvent.

(iv) Expansion

When a liquid is heated the particles acquire more kinetic energy and move further away from each other. This makes the liquid expand.

(v) Evaporation

The particles in a liquid have different amounts of energy. Some have less kinetic energy than others and move more slowly than them. Others have more kinetic energy and move more quickly. At the surface of the liquid the particles with the largest amount of kinetic energy move so fast that they can pull away from the force of attraction that exists between them and other liquid particles and escape into the air as vapour.

If a liquid is warmed the energy of all its particles is increased and more particles acquire enough energy to escape from the surface. This increases the rate of evaporation.

(vi) Boiling

If a liquid is heated strongly enough the particles with the most kinetic energy separate from the particles with less kinetic energy inside the liquid. When this happens the fast-moving particles form bubbles in the liquid. As the bubbles contain fewer particles than the surrounding liquid they are less dense than the liquid and rise to the liquid surface where they burst and the fast-moving particles are released into the air.

QUESTION

20 Describe how evaporating and boiling are (a) similar, (b) different by considering how the particles behave during each process.

C THE GASEOUS STATE

In gases the particles have a great deal of kinetic energy and easily overcome their attraction for each other and are spread out. They move freely and quickly in all directions as Figure 1.11 shows,

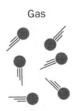

Gas

Figure 1.11   Particles moving in a gas

(i) Expansion

When a gas is heated its particles acquire more kinetic energy. This makes them move further away from each other and if the gas is free to expand its volume increases.

(ii) Gas pressure

The particles in a gas push on the surfaces around them. The pushing force that they exert is called **gas pressure**. If the temperature of the gas is increased but its volume is not allowed to increase its particles move more quickly and strike the surfaces more frequently. This increases the gas pressure. If the temperature of the gas is lowered the particles move more slowly and the pressure is reduced.

A gas can be squashed into a small space. If this is done without raising the temperature of the gas the pressure still rises. This happens because the particles are still moving at the same speed but because they have less space to move about in they strike the surfaces of the container more frequently.

(iii) Diffusion

Diffusion occurs in gases, liquids and solids. It occurs in the same way in each state. Diffusion is quickest in gases and slowest in solids. Diffusion in a solid is very slow and can only be seen by examining the inside of the solid.

For the diffusion of a substance to take place its particles must be in a region of high concentration and free to move to a region of lower concentration. They do this by moving between the other particles in their surroundings. The number of particles present in the region of high concentration, low concentration and the region in between can be represented as a line called a diffusion gradient as shown in Figure 1.12.

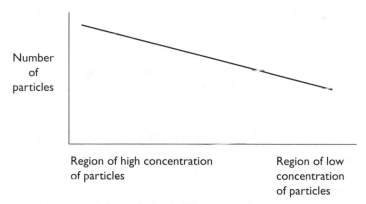

Figure 1.12   A diffusion gradient

If the gradient is steep the rate of diffusion will be fast but if the gradient is shallow the rate of diffusion will be slow. When the particles are completely dispersed there is no difference in concentration and the diffusion gradient ceases to exist.

(iv) Factors affecting diffusion

The rate of diffusion is increased when the temperature of the diffusing substances is increased. It is decreased when the temperature of the diffusing substances is decreased.

Smaller particles diffuse more quickly then larger particles.

In gases, the size of the spaces between the particles affects the rate of diffusion. When the spaces are large the particles can move more freely and diffusion takes place more quickly. When the pressure of a gas is increased the spaces become smaller and the particles collide more frequently. These collisions slow down the rate of diffusion. When the pressure is decreased the spaces between the particles is increased and the rate of diffusion increases too.

QUESTIONS

21 Look back at the diffusion of ammonia and hydrogen chloride gas (see p. 7) and use the kinetic theory to explain why the white ammonium chloride does not form half way down the glass tube.

22 Explain Boyle's Law (see p. 7) in terms of the kinetic theory.

23 Explain Charles' Law (see p. 8) in terms of the kinetic theory.

(v) Condensation

If a gas is cooled down enough the particles lose so much kinetic energy that they slow down, move closer together until they start being attracted to each other and slide over each other and form a liquid.

(vi) Sublimation

This only occurs in a few substances. When the particles in a few solids receive enough energy from a heat source they move so quickly that they separate completely.

When the particles in a few gases are cooled they slow down so quickly that they become fixed together and form a solid.

QUESTION

24 Compare evaporation, condensation and sublimation using the kinetic theory.

1.12 Density

The density of a substance is found by **dividing its mass by its volume**. The mass of a substance is the amount of **matter** it contains. The amount of matter in a volume of a substance depends on the number of particles in that volume and their mass. A volume of a substance with a large number of particles in it is denser than the same volume of substance with fewer particles of the same mass. A volume of a substance with particles with a large mass has a greater density than the same volume of a substance with the same number of particles with a smaller mass.

1.13 Summary

● There are three states of matter – solid, liquid and gas (see p. 1).
● The states can change through melting, freezing, evaporation, boiling, condensation (see pp. 4–5).

- The relationship between the temperature, pressure and volume of a gas is described by the gas law (see p. 9).
- The kinetic theory explains the states of matter in terms of tiny moving particles (see p.11).
- Dissolving can be explained by the kinetic theory (see p. 13).
- Diffusion can occur in solids, liquids and gases (see p. 15).
- Density can be explained by the kinetic theory (see p. 16).

■ ⋈ 2 The Earth and the atmosphere

Objectives

When you have completed this chapter you should be able to:
- Explain how the **Solar system** formed
- Describe the changes that have taken place in the **Earth's atmosphere**
- Describe the **structure** of the Earth
- Distinguish between **igneous**, **sedimentary** and **metamorphic rock**
- Explain how **fossils** form
- Distinguish between **weathering** and **erosion**
- Describe the **rock cycle**
- Describe the evidence for **continental drift**
- Explain the **Plate Tectonic theory**.

2.1 How the Solar system formed

The first elements to form after the Big Bang were **hydrogen** and **helium**. The force of gravity pulled them together into clouds called **nebulae**. Inside a nebula the hydrogen and helium atoms continued to be pulled together by the force of gravity and formed spheres of gas. As the hydrogen atoms gathered together at the centre of a gas sphere they became squashed and hot. The high temperature and pressure produced a fusion reaction in which some hydrogen atoms were converted to helium atoms and a large amount of energy was released. The energy passed from the centre of the gas sphere to its surface and escaped as light and heat. The gas sphere became a **star**.

When a star uses up its fuel of hydrogen atoms it makes other elements with larger atoms such as carbon or iron. When a star has finished this process it either releases them as dust as it fades out or hurls them into space as it explodes as a super nova.

Clouds of hydrogen and helium and the dust made from stars gathered to make more stars which in time produced even more elements in the cosmic dust.

About 5000 million years ago a cloud of gas and dust formed the Sun and the planets, moons, asteroids and comets of the Solar system.

1 How is light energy produced by hydrogen atoms in a gas cloud?
2 What is the connection between a pebble and a super nova that occurred millions of years ago?

As the cloud of gas and dust swirled round the Sun the particles of dust joined together to form lumps of rock. These joined together to form larger pieces. Eventually they formed the rocky planets Mercury, Venus, Earth, Mars and Pluto. Four planets are formed mainly of gas and liquids. They are Jupiter, Saturn, Uranus and Neptune.

When the Earth first formed it was a huge globe of molten rock. The heat was generated by the materials being squashed by the gravitational forces at the centre and also by the decay of radioactive materials (see p. 22).

2.2 The Earth's atmosphere

Hydrogen and helium gathered around the planet to form its first or primary atmosphere. The atoms of these gases are very small and they were driven away by streams of charged particles rushing through space as they escaped from the Sun. These charged particles make the solar wind which still rushes past the Earth today.

In time, the surface of the Earth cooled but the heat beneath it caused widespread volcanic eruptions. The lava flowing from the volcanoes formed igneous rocks.The gases which escaped from the volcanoes – carbon monoxide, carbon dioxide, ammonia, hydrogen sulphide, methane and water vapour – formed the secondary atmosphere which was not blown away by the solar wind.

As conditions at the surface became cooler still the water vapour in the atmosphere condensed to form clouds which in turn produced rain. The water flowed over the rocks dissolving some substances and breaking up some of the rocks into fragments. The water collected in the large depressions in the Earth's crust and formed seas and oceans. The dissolved substances made the sea water salty and as the water splashed down rivers and formed large ocean surfaces ammonia and carbon dioxide from the atmosphere dissolved in it too.

Gradually the atmosphere began to change. Ultra violet light could penetrate all the way through the atmosphere and break up molecules of water in water vapour to form molecules of hydrogen and oxygen gas. Ammonia molecules were also broken up to form molecules of nitrogen and hydrogen. The winds rushing through the huge clouds produced electrostatic charges on the particles in the clouds which in turn produced flashes of lightning (Figure 2.1). The lightning also caused the decomposition of water and ammonia molecules.

The lighter molecules of hydrogen drifted off into space while the heavier molecules of oxygen and nitrogen remained in the atmosphere. The first forms of life used the energy locked in chemical compounds in the seas and oceans. Later forms, algae and green plants used energy from sunlight to make food in the process called **photosynthesis**. Oxygen is also produced in this process and as

Figure 2.1 A lightning flash

Component	%
Nitrogen	78
Oxygen	21
Noble gases	0.9
Carbon dioxide	0.03
Water vapour	Variable

Table 2.1 The composition of the atmosphere

the algae and the plants thrived the amount of oxygen in the atmosphere increased.

Planktonic animals and many kinds of invertebrates evolved which took in some of the carbon dioxide dissolved in the sea water. They used the carbon dioxide to make calcium carbonate – a hard substance which they used to make their protective shells. When many of these types of organisms died their shells formed rocks.

Plants and animals use up oxygen and produce carbon dioxide in the process of respiration. Today the composition of the atmosphere is shown in Table 2.1.

QUESTION

3 How has the composition of the Earth's atmosphere changed over time?

2.3 The structure of the Earth

When an earthquake takes place two kinds of waves called **primary** and **secondary waves** are produced which pass through the planet. The primary waves can pass through all the materials they encounter but their direction is bent as they pass from one material to another. The secondary waves cannot pass through liquids and when they reach the surface of a liquid they are deflected back to the surface.

The structure of the Earth has been worked out largely by the study of where the primary and secondary waves from an earthquake have revisited the surface of other parts of the planet From this information the structure of the Earth is believed to be as shown in Figure 2.2.

(i) The core

The core has a diamcter of 6378 km and is composed of iron and nickel. It is divided into an inner and outer core.

(ii) The inner core

The inner core is a ball of solid rock with a diameter of 2530 km. It is at a temperature of over 3000°C but the great pressure due to the other material in the Earth pushing down on it keeps it in the solid state.

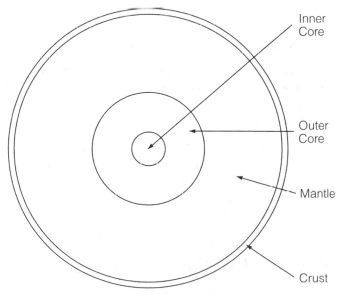

Figure 2.2 The structure of the Earth

(iii) The outer core

The outer core is at a high temperature but the pressure is lower due to less material pushing down. This reduction in pressure allows the iron and nickel to exist in the liquid state. It is believed that as the planet rotates on its axis this liquid also moves and generates the magnetic field which passes through the Earth and into a region of space around the planet.

(iv) The mantle

The mantle is a layer of hot rock. It is 2900 km thick and parts of it are semi-solid and can flow. Parts of the mantle which flow beneath the crust cause the movement of the crustal plates. The temperature of the mantle is 600–1200°C. The heat is caused by the radioactive decay of isotopes of thorium potassium and uranium.

QUESTIONS

    4 How thick is the core layer that separates the inner core from the mantle?
    5 How is the mantle different from the core?

(v) The crust

The crust is the layer of the rock which covers the surface of the planet. It is less dense that the rock in the mantle and so floats on top of it. The core and the mantle formed 4500 million years ago when the planet came into being but the crust is much younger. There are two types of crust. They are the continental crust which formed 3–500 million years ago and the oceanic crust which began forming about 200 million years ago. The temperature of the rocks in the crust range from 10–600°C.

(vi) Continental crust

This is about 35 km thick. It is composed mainly of granite, an igneous rock but it also includes sedimentary rocks and metamorphic rocks.

(vii) Oceanic crust

This is about 8 km thick. It is composed mainly of basalt. Continental crust is less dense than oceanic crust and floats on top of it.

QUESTION

    6 In what ways do the two types of crust differ?

2.4 The types of rock in the Earth's crust

There are three types of rock in the Earth's crust. They are **igneous, sedimentary** and **metamorphic** rocks.

(i) Igneous rocks

Igneous rock forms from magma. This is molten rock from the mantle. Magma is less dense than the crust and so rises through it. There are two kinds of igneous rock:

(a) Intrusive igneous rock (Figure 2.3)

This forms when the magma fails to reach the surface of the crust. The rock may form a dome-like structure called a **batholith** or it may fill a vertical crack in the rock and form a **dyke**. The magma may also flow between layers of sedimentary rock (see p. 24) and form a **sill**.

When the **magma** fails to reach the surface it cools slowly and forms a rock with large crystals. This rock is called **granite**. The rock may take many years to cool down depending on the amount of magma that has collected in the crust.

The surface of granite can be polished to give it an attractive finish due to its large crystal structure. Granite is used to build the entrances and fronts to buildings such as libraries and museums. It is a hard rock and is used in making concrete and in the foundation of roads.

(b) Extrusive igneous rock

This forms when the magma reaches the surface. The rock pours out of a volcano and cools down quickly. It forms a rock composed of small crystals. This rock is called **basalt**. It is also a hard rock and is used in making concrete and road foundations. It is also used in nuclear reactor shields.

QUESTION

    7 In what ways are the two kinds of igneous rock different?

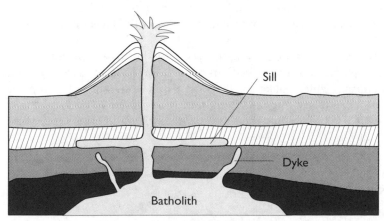

Figure 2.3   Batholith, dyke and sill

(ii) Sedimentary rock

Three-quarters of the continental crust is covered by sedimentary rocks. They take millions of years to form. There are three kinds of sedimentary rock:

(a) Clastic sedimentary rock

This is produced by weathering and erosion of other rocks. Examples are **sandstone**, **shale** and **mudstone**. Sandstone is used for making buildings and concrete.

(b) Biogenic sedimentary rock

This is produced by the deposition of materials from living things. Most **limestones** are formed from the shells of invertebrate animals such as corals and molluscs which lived in ancient seas. Limestone is used for making buildings and in the extraction of iron from its ore.

Chalk is formed from the tiny shells or tests of microscopic animals which lived in the plankton of ancient seas.

Coal is formed from the fossilisation of swamp plants that lived 355 million years ago. Coal is used as a fuel.

(c) Chemical sedimentary rock

Rock salt is formed from salt which has been left behind when the sea water evaporated and the salt was compressed which made it into hard rock. Rock salt is spread on the roads in winter to prevent the formation of ice.

QUESTION

8 How can you distinguish between the three types of sedimentary rock?

(iii) Sedimentary rocks and fossils

Sedimentary rocks are laid down in layers. Generally the deeper layers are older than the ones above but in mountain formation the layers may be folded over each other.

Fossils form in sedimentary rocks. Most of the fossils are formed from the remains of plants and animals although some may be formed from animal tracks. The chemicals which make up the organism's body are either replaced with minerals or crystals of minerals form in the body tissues. The fossils present in a rock can be used to identify the time at which the sedimentary rock was formed (Table 2.2).

QUESTION

9 A rock has a trilobite fossil in it and another rock has a sea urchin fossil in it. What information does this give you about the rocks?

MYA	Period	Fossil
2	Quarternary	mammoth
65	Tertiary	snail
136	Cretaceous	ammonite
190	Jurassic	sea urchin
225	Triassic	fish
280	Permian	lampshell
355	Carboniferous	coral
395	Devonian	fish
440	Silurian	graptolite
500	Ordovician	lampshell
570	Cambrian	trilobite

Note: MYA = millions of years ago.

Table 2.2   Table of fossils from different ages of the Earth

(iv) The formation of sedimentary rock

When a rock in the Earth's crust is exposed at the crust's surface it is broken down by the processes of weathering and erosion.

(a) Chemical weathering

In this process the chemical composition of the rock is changed. This is largely produced by the action of rain water which contains dissolved carbon dioxide, nitrogen dioxide and sulphur dioxide. These dissolved gases make an acidic solution which dissolves compounds in the rock such as calcium carbonate. The parts of the rock which do not dissolve are left without the support of the parts that have dissolved. This makes the rock weaker and easier to break up by other forms of weathering and erosion.

(b) Physical weathering (Figure 2.4)

There are two kinds of physical weathering:

Exfoliation

Rocks are made from a mixture of different minerals. A rock at the surface of the crust is warmed and cooled by the weather. When the rock is warmed the different minerals expand at different rates from each other. This produces strain forces in the rocks. Similar forces are produced when a rock is cooled due to the way the different minerals contract at different rates. These forces which are set up in the rock as it is warmed and cooled cause parts of the rock to break up.

Freezing and thawing

During a shower the cracks in a rock fill up with water. If the temperature falls and the water freezes ice is produced. Water expands as it freezes and pushes on

Figure 2.4 Weathered rocks

the rocky sides of the crack. The size of the force may cause pieces of rock to break off.

(c) Biological weathering

Trees and shrubs are perennial plants. This means that they live for many years and increase in size every year. If the tree or shrub roots are growing in a crack in a rock their increase in size exerts a force on the rock which causes parts of it to break up.

QUESTION

10 How may a rock be broken up by weathering?

(d) Erosion

Erosion is the wearing away of the rock by water, ice and solid particles in the wind:

Water and ice

As water in a river rushes over the rocky river bed it carries rocks and pebbles which chip away at the rocky surface of the bed. In time, a river may erode way so much rock that it creates a deep sided valley called a gorge.

In mountainous and polar regions glaciers move over rocky surfaces and break them up.

Wind

When solid particles such as sand grains are carried in a strong wind they hit rocky surfaces with such force that fragments break off.

QUESTION

    11   What is the difference between weathering and erosion?

(e) Transport of weathered and eroded material

The fragments of rock produced by weathering and erosion are carried away by water in rivers or the waves of the sea. The fragments eroded by a glacier are carried along with the ice and may cause further erosion as they rub against the rocky surface. The smallest particles produced by wind erosion may travel in the air and cause more erosion.

(f) Deposition

A river may carry a range of sizes of rocky fragments but as it slows down the fragments are deposited in order of size. The largest fragments settle down first. They form a sedimentary rock called **conglomerate**. Sand particles settle out next and form **sandstone** while the smallest particles which may be clay and mud settle out last and form **shale or mudstone**.

(v) How rock particles form new rock

When fragments of rock are deposited they form layers. As the layers increase in thickness the pressure on the fragments also increases and they are forced closer together. The water which deposited the fragments is a solution of minerals and as it is squeezed out from between the fragments the minerals come out of solution and coat the fragments and cement them together.

(vi) Metamorphic rocks

Metamorphic means 'change of form'. A metamorphic rock is one that has undergone a change of form.

There are two kinds of metamorphic rock:

(a) Contact metamorphic rock

This rock is produced when rocks in the crust are heated by igneous rocks as they pass through the crust towards the surface. **Marble** is produced when limestone is heated in this way. Marble is used for the interior of prestigious buildings such as libraries, museums and banks. **Slate** is produced when shale is heated by igneous rock. Slate forms thin sheets of rock which are impervious to water and has in the past been extensively used as a roofing material.

(b) Regional metamorphic rock

This rock is produced by the great pressures which are produced when the layers of rock become piled up during the formation of mountains. Examples of rocks which form in this way are **gneiss** (pronounced 'nice') which is produced from limestone and **schist** which has bands of interlocking crystals and is made from mudstone.

QUESTIONS

12 What is a metamorphic rock?
13 How can metamorphic rocks be formed?
14 In what ways are marble and gneiss (a) similar, (b) different?

2.5 The rock cycle

James Hutton (1726–97) observed rock formations in the landscape and put forward the idea that rocky material changed from one form into another by processes such as erosion and deposition. Today his ideas have been expanded into the concept of the rock cycle (Figure 2.5).

QUESTION

15 Using Figure 2.5 (a) describe the history of the elements in a metamorphic rock from the time they were in the magma in the Earth's crust. (b) Predict the future of the elements in the metamorphic rock.

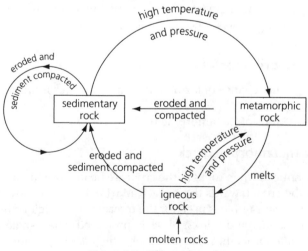

Figure 2.5 The rock cycle

2.6 Continental drift

If you look at the map of the world in Figure 2.6 you can see that the north west side of Africa looks as if it could have fitted in the space between North and South America.

Further studies suggest that the continents were once close together and have drifted apart.

These studies include:

(i) Fossil evidence

South America, Africa, India and Antarctica shared the same fossils in the distant past but in more recent geological times their fossil records differ. This suggests that at one time the land in the different continents were so close together that they were populated by the same organisms. At a later time the continents moved apart and the organisms evolved in different ways and left different fossils behind.

(ii) Magnetic record in the rocks

When an igneous rock is newly formed the magnetic elements in it are free to move and arrange themselves in a north–south direction in the Earth's magnetic field. The direction of the magnetic north and south is changing very slightly all the time and periodically when their positions have changed so much maps have to be reprinted with the new information. Sometimes the positions of the poles completely reverse. The position of the magnetic materials in rocks which formed in South America, Africa, India and Australia about 200 million years ago are similar. This shows that they formed in a similar place on the Earth's surface near the geographic south pole (see Figure 2.7).

Figure 2.7 shows the position of the land 250 million years ago. This huge land mass is called 'Pangea', meaning 'all earth'. Before Pangea formed there were other smaller land masses. Some of these were the size of continents today but they were a different shape.

The original concept of continental drift was put forward by Alfred Wegener (1880–1930). He believed that as the continents were made mostly of granite which is lighter in weight than basalt in the oceanic crust, they floated on the basalt and gradually drifted apart. Today the movement of the continents is explained by the **Plate Tectonic theory**.

QUESTION

16 What is the evidence to suggest that the continents drifted apart?

2.7 Plate Tectonic theory

The Plate Tectonic theory explains the movement of the continents in the following way:

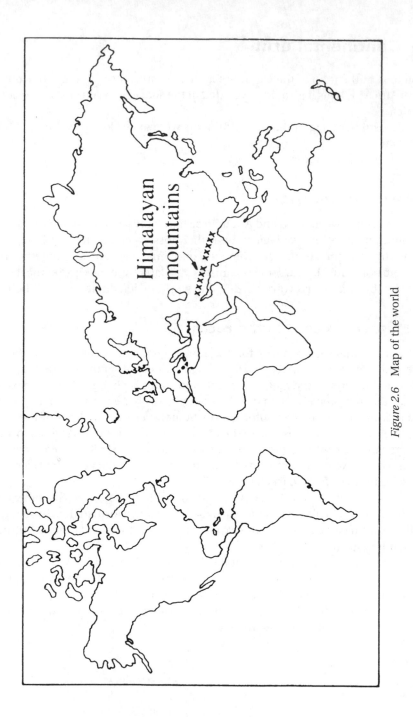

Figure 2.6 Map of the world

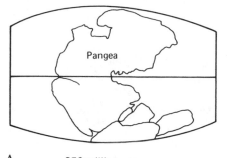

A   250 million years ago

Figure 2.7   Land 250 million years ago

The Earth's crust floats on the mantle and is divided into sections called plates, There are six large plates and a few smaller ones (see Figure 2.8); most of the plates have a continent on them with some oceanic crust. A few plates are composed of oceanic crust.

QUESTION

17   Compare Figures 2.7 and 2.8. How has the land mass on the planet changed in 250 million years?

The plates are in constant, very slow motion across the surface of the planet. The power for this movement comes from the convection currents in the mantle which in turn have been set up by the heat inside the Earth.

As a plate moves its edges or margins touch the edges or margins of neighbouring plates. The movements of the plates may cause the margins of the two plates to behave in one of the following ways:

(i) Slide by each other

As the plate edges rub together **earthquakes** develop. The San Andreas Fault on the coast of California is a place where the Pacific and North American Plates are sliding by each other.

(ii) Move towards each other

When an oceanic plate and a continental plate move towards each other – for example, the Nazca Plate and the South American plate – the oceanic crust dips under the continental plate in a process called **subduction** (Figure 2.9). This produces an ocean trench and fold mountains and volcanoes on the continental crust.

When, two continental plates – such as the Indo-Australian Plate and the Eurasian Plate – move towards each other neither plate is subducted and mountain ranges like the Hymalayas are created.

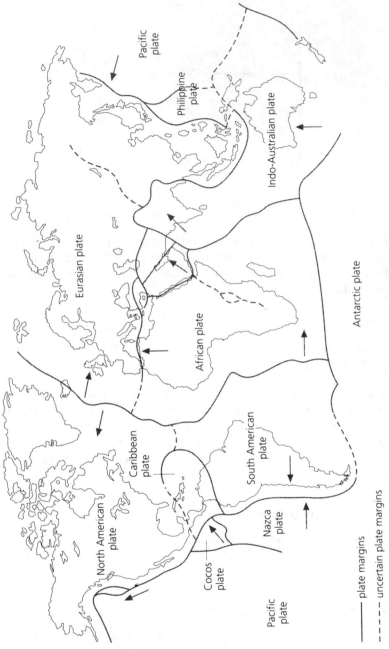

Pacific
plate

Philippine
plate

Indo-Australian plate

Eurasian plate

Antarctic plate

African plate

North American
plate

Caribbean
plate

South American
plate

Cocos
plate

Nazca
plate

Pacific
plate

——— plate margins
- - - - uncertain plate margins

Figure 2.8 Map of the world showing the continental plates

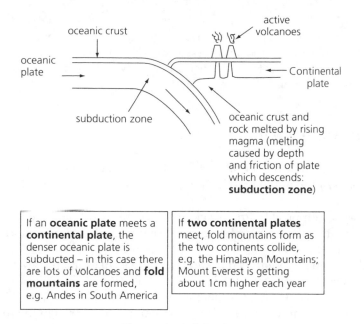

Figure 2.9   Subduction

(iii) Move away from each other

When two oceanic plates move away from each other – as in the mid-Atlantic – a ridge forms from the basalt which is produced by the magma rising up through the gap produced by the parting plate margins (Figure 2.10).

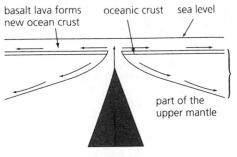

Figure 2.10   Formation of an oceanic ridge

When two continental plate margins move apart a rift valley is created. The Great Rift valley in East Africa has formed in this way.

QUESTIONS

18 Explain what you see in Figures 2.8, 2.9 and 2.10.
19 What is the difference between an oceanic trench and an oceanic crust?

20 When the rocks on either side of an oceanic ridge are examined some are found to have magnetic materials arranged in exactly the opposite way to the magnetic materials in other rocks. Look at the information about the magnetic record of the rocks on p. 29 and suggest how this difference in the rocks may have occurred.

2.8 Summary

- The Solar system formed from a cloud of gas and dust (see p. 18).
- The composition of the Earth's atmosphere has changed since it first formed (see p. 19).
- The Earth is composed of three regions – the core, mantle and crust (see p. 21).
- There are three types of rock – igneous, sedimentary and metamorphic (see pp. 22 and 28).
- Fossils form in sedimentary rocks and can be used of estimate the age of a sedimentary rock (see p. 24).
- Rocks are broken down by weathering and erosion (see pp. 25–7).
- The material from which rocks are made changes as shown in the rock cycle (see p. 28).
- There is evidence to suggest that the continents are drifting apart (see p. 29).
- The Plate Tectonic theory is used to explain continental drift (see p. 29).

3 Investigating substances

Objectives

When you have completed this chapter you should be able to:
- Identify **laboratory apparatus**
- Distinguish between **elements**, **mixtures** and **compounds**
- Describe the processes of **decanting, filtering** and **centrifuging**
- Distinguish between **evaporating** and **crystallisation**
- Describe simple **distillation and chromatography**
- Describe how **immiscible liquids** can be separated
- Describe how **miscible liquids** can be separated by fractional distillation
- Describe how solids can be separated by **sublimation**
- Understand how the purity of a substance can be **checked**.

The science of chemistry developed from the an activity called alchemy. This was practised for over two thousand years and two of its aims were to discover a way to turn cheap metal into gold and to make a medicine which could prolong life indefinitely. Neither aims were realised but from the work of alchemists some knowledge of matter was discovered. Some pieces of apparatus were invented and some techniques of separating mixtures were developed.

3.1 Simple laboratory apparatus

A wide range of chemical investigations can be performed using a small number of pieces of apparatus. The most frequently used apparatus are illustrated in Figure 3.1. In other parts of the book they and others not shown here are illustrated diagramatically as line drawings for simplicity.

QUESTIONS

1 How can you distinguish between a test tube and a boiling tube?
2 Which pieces of apparatus give the (a) most accurate readings of volumes of liquids, (b) least accurate readings of volumes of liquids?
3 How are the different types of Bunsen burner flame produced?
4 How do you think the size of the flame is controlled? (*Hint*: the gas released from the tap is the fuel.)
5 When would you use (a) a gauze, (b) a pipe clay triangle? Explain your answer.

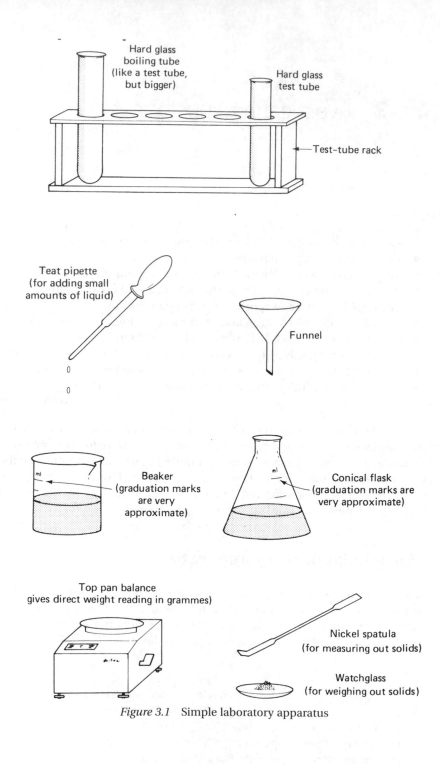

Figure 3.1 Simple laboratory apparatus

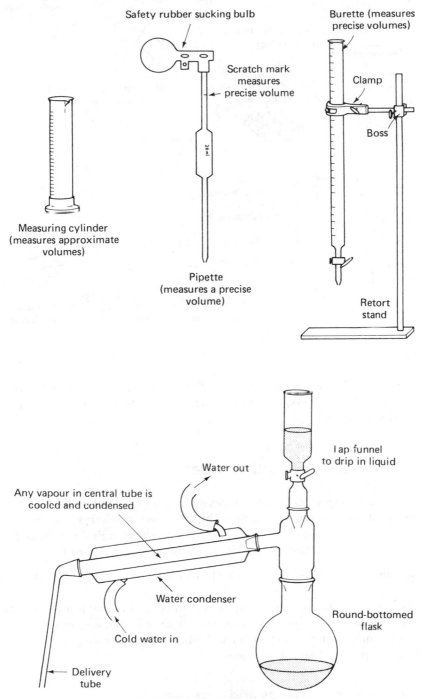

Safety rubber sucking bulb

Scratch mark measures precise volume

2 ml

Measuring cylinder (measures approximate volumes)

Pipette (measures a precise volume)

Burette (measures precise volumes)

Clamp

Boss

Retort stand

Tap funnel to drip in liquid

Water out

Any vapour in central tube is cooled and condensed

Water condenser

Cold water in

Delivery tube

Round-bottomed flask

Figure 3.1   Continued

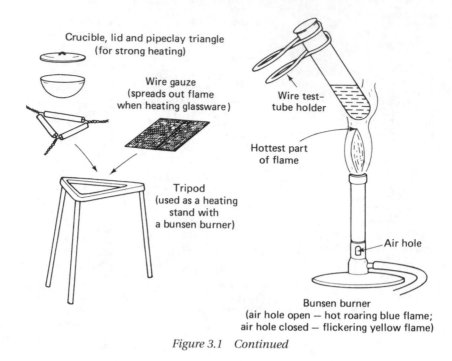

Crucible, lid and pipeclay triangle
(for strong heating)

Wire gauze
(spreads out flame
when heating glassware)

Wire test–
tube holder

Hottest part
of flame

Tripod
(used as a heating
stand with
a bunsen burner)

Air hole

Bunsen burner
(air hole open — hot roaring blue flame;
air hole closed — flickering yellow flame)

Figure 3.1   Continued

All practical work must take place under the supervision of a qualified teacher or lecturer and all safety procedures in operation at the school or college must be followed at all times.

3.2 The nature of substances

Alchemists investigated many substances in their work and found that some substances could be easily separated by techniques such as filtering and evaporating while other substances could only be obtained by chemical reactions. From their work substances could be placed in one of three main groups:

● **Elements** – substances which cannot be broken down into simpler substances (see also p. 49).
● **Mixtures** – substances which can be separated into two or more substances using physical processes such as filtering, boiling and condensing.
● **Compounds** – substances which can be broken down into other substances by taking part in chemical reactions.

(i) Comparing mixtures and compounds

Every element has specific physical and chemical properties. When two elements form a mixture each element retains its physical and chemical properties. No chemical reactions take place between them. No heat is taken in or given out and the amounts of the elements in the mixture can vary.

When two elements form a compound they no longer retain their physical and chemical properties. The compound exhibits its own physical and chemical properties, which are different from those of the elements that form it. A chemical reaction takes place between the elements when the compound forms and during the reaction heat may be given out or taken in. The percentage of each element in a compound is fixed. They are combined in certain proportions which cannot be varied.

QUESTIONS

6   Iron and sulphur are found in a yellow-grey form and in a dark grey form. The element iron is magnetic. When a magnet is brought close to the yellow-grey form the grey particles stick to it. Nothing happens when the magnet is brought close to the dark grey form – no force of attraction is felt. Explain these observations.

7   Is it possible to have a mixture of (a) elements and compounds, (b) compounds and compounds? (*Hint*: think about iron filings, salt and sugar.)

3.3 Separating substances

The most common mixtures which are separated in a chemistry laboratory are mixtures of solids and liquids and mixtures of liquids.

(a) Separating liquids and insoluble solids

When some solids and liquids are mixed the solid does not dissolve in the liquid. It is said to be insoluble in the liquid.

Decanting

When an insoluble solid is mixed with a liquid it may form a suspension of particles floating in the liquid. In time the particles may settle at the bottom of the container and form a sediment. Larger particles form a sediment more quickly than smaller particles. A clear liquid is left above the sediment which can be carefully poured or decanted leaving the **sediment** (also called the residue) behind.

Filtering

A mixture of a liquid and an insoluble solid can be separated without waiting for the sediment to settle by using a process called filtration (see Figure 3.2).

The filter paper has tiny holes called pores through which the liquid can pass. The solid particles are too large to pass through the pores and stay behind in the filter paper and form the residue. The liquid which passes through

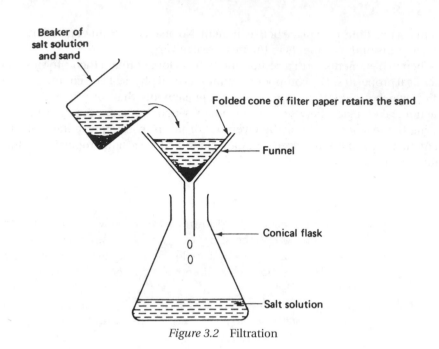

Beaker of salt solution and sand

Folded cone of filter paper retains the sand

Funnel

Conical flask

Salt solution

Figure 3.2 Filtration

the filter paper is called the **filtrate**. It may be a pure liquid or it may be a solution as the particles of solute (see p. 13) are small enough to pass through the pores.

Centrifuging

Some insoluble solid particles are so small that they do not settle to form a sediment but remain floating in the liquid and form a suspension which makes the liquid cloudy. If the mixture is filtered the particles cannot pass through the pores in the filter paper but they are small enough to block them and stop the filtration process.

A centrifuge (Figure 3.3) is used to separate the solid in suspension from the liquid. It is a machine which spins test tubes of the mixture very quickly. This action forces the solid particles to the bottom of the test tube where they form a residue. When the test tubes are removed from the centrifuge the solid and liquid can be separated by decanting the liquid.

QUESTION

8 Why are there three different ways of separating insoluble solids from liquids?

(b) Separating a soluble solid from the solvent

These methods are used to separate a soluble solid from its solution. A substance which dissolves in a liquid is called the solute. The liquid in which the solute dissolves is called the solvent.

Figure 3.3   A centrifuge

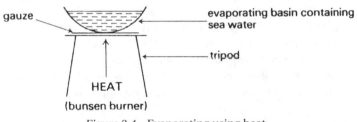

Figure 3.4   Evaporating using heat

Evaporation

When a liquid surface is exposed to the air some of the particles from which it is formed pass into the air due to evaporation (see p. 13). This process can be used to separate the solvent from the solid solute because the particles of the solid solute cannot escape into the air.

The mixture to be separated by evaporation is poured into an evaporating dish. This may then be left for the liquid to evaporate at room temperature or it may be heated as shown in Figure 3.4.

As some liquid evaporates the concentration of the solute in the liquid remaining in the evaporating dish increases. Eventually the solution becomes saturated. This means that there is not enough liquid present to be able to dissolve all the solid. It can hold no more of the solid in solution. Further evaporation of a saturated solution leads to the production of crystals of the solid. If the evaporating dish is being heated the heat may cause the crystals to decompose. If the crystal form of the solid is required the crystallisation process is used, as described in the next section.

Crystallisation

This process is similar to the evaporation process using heat until the saturated solution stage is reached. When this stage is reached the heat is removed from the mixture and the mixture is allowed to cool. Evaporation continues during this process and also when the mixture has reached room temperature. The solid left behind when evaporation is complete is in crystalline form.

(c) Separating the solvent from the soluble solid

Distillation

In both evaporation and crystallisation the solvent is allowed to escape into the air and is not collected. In distillation the solvent, once it has been turned into a vapour (gas), is allowed to condense (turn back into a liquid) away from the solid so both solid and solvent may be collected (Figure 3.5).

Figure 3.5 shows distillation.

The solution in the flask is heated and allowed to boil. This produces rapid evaporation of the solvent which rises to the top of the flask in gaseous form. The thermometer shows the boiling point of the liquid. The gas leaves the flask and enters the central tube of the condenser. There is a glass chamber called the water jacket around the central tube. A current of cold water is directed through the water jacket. It flows in the opposite direction to the gas in the central tube. The gas loses heat to the cold water and cools down. As the temperature of the substance falls below its boiling point it condenses on the walls of the central tube to form a liquid which flows down the tube to the beaker for collection. The solid remains in the flask after all the liquid has been boiled away.

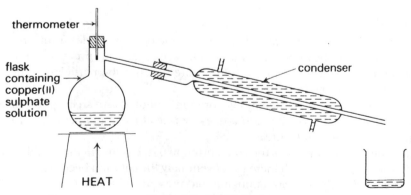

Figure 3.5 Distillation

9 Compare the process of evaporation with distillation. How are they (a) similar, (b) different?

(ii) Separating several solids from a solution

Different solids have different solubilities in a solution. When they are in solution they can also be absorbed by special paper, called chromatography paper but the paper's power to absorb a solid is different for each solid. These differences in physical properties of the solids are used to separate them in a process called **chromatography**.

A common example of chromatography is in the separation of the different coloured solids that are mixed together to make a dye or an ink. The solids are separated by placing a drop of the solution near the end of a piece of chromatography paper then dipping the end into a solvent and allowing the solvent to move up through the paper, as Figure 3.6 shows.

When the solvent reaches the drop of ink the coloured solids dissolve in it and are carried further up the paper. The different solids settle out on the paper at different places due to the differences in their solubilities and the paper's power to absorb them.

Chromatography can also be used to separate solids which are not coloured. When the solvent has reached the top of the paper the paper is sprayed with a chemical called a locating agent which makes the solids take up a colour so that they can be seen.

QUESTIONS

10 In Figure 3.6b, which solid is least soluble and is most readily absorbed by the paper?
11 Explain why solid B has settled in a position which is different from solids A and C.
12 How could the contents of a second ink be compared with A, B and C in Figure 3.6b?

(iii) Separating two solids from a mixture

A few solid substances sublime. **Sublimation** means that the solid changes directly into a gas when it is heated and the gas turns directly into a solid when it is cooled.

If a mixture contains a solid with a high boiling point and one that sublimes at a lower temperature it can be heated to separate the two solids. A mixture of salt and iodine crystals is an example of a mixture which can be separated in this way. When the mixture is heated the iodine solid changes into a gas while the salt remains a solid. The heated mixture is covered with a cooled filter funnel as Figure 3.7 shows.

The iodine vapour is cooled by the filter funnel and condenses into a solid on its sides. The salt remains in the evaporating dish.

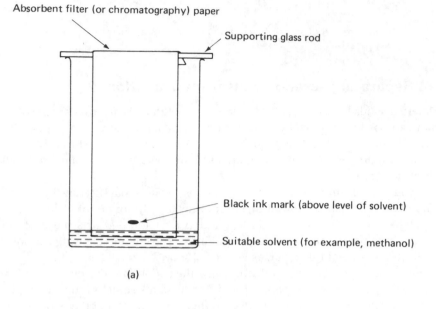

Absorbent filter (or chromatography) paper

Supporting glass rod

Black ink mark (above level of solvent)

Suitable solvent (for example, methanol)

(a)

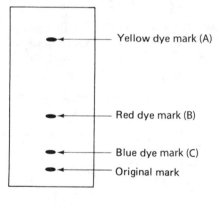

Yellow dye mark (A)

Red dye mark (B)

Blue dye mark (C)

Original mark

(b)

Figure 3.6   Simple chromatography

QUESTION

13   Why is separation by sublimation not a widely used technique (see p. 16)?

(iv) Separating liquids

When two liquids are poured together they may mix or they may not.

(a) Immiscible liquids

Liquids which do not mix are described as being **immiscible**. When water and oil are poured together they separate into layers. They are immiscible.

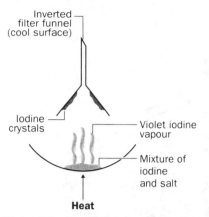

Figure 3.7 Separation by sublimation

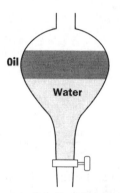

Figure 3.8 Two immiscible liquids in a separating funnel

Immiscible liquids can be separated by using a separating funnel (see Figure 3.8).

When the two liquids have settled to form layers the tap at the bottom of the funnel is opened and the lower liquid is allowed to escape into a beaker. When the lower liquid layer has left the funnel the tap is closed and the remaining liquid can be poured out into a second beaker.

(b) Miscible liquids

Liquids which mix when they are poured together are described as being **miscible**. You cannot tell there is more than one liquid by looking at the mixture. Water and ethanol are examples of two liquids which are miscible.

Miscible liquids have different boiling points and each liquid in the mixture keeps its own boiling point. This difference in one of their physical properties is used to separate them in a process called fractional distillation. The apparatus for fractional distillation is shown in Figure 3.9.

When the mixture of liqiuds in the flask is heated the liquid with the lowest boiling point boils first. At that temperature the speed of evaporation of the

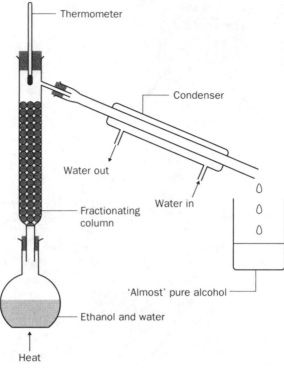

Figure 3.9 Fractional distillation

second liquid is also increased and some of it may escape into the fractionating column too. The liquid with the lower boiling point rises to the top of the fractionating column and enters the condenser but the liquid with the higher boiling point condenses on the beads in the column and flows back to the liquid in the flask. This happens because the column is at the temperature of the boiling point of the liquid with the lower boiling point. This is below the boiling point of the second liquid so any gas that escaped by evaporation from the second liquid is cooled.

As the liquids are both colourless the thermometer is used to tell when they have separated. The separation of the two liquids is complete when the thermometer starts to rise for a second time. This means that all the liquid with the lower boiling point has escaped and the rise is due to the second liquid which is now starting to boil.

Fractional distillation can be used to separate a mixture of several liquids. It is used in the separation of gases from liquid air and in the separation of components in oil.

QUESTIONS

14   Water boils at 100°C and ethanol boils at 78°C.
(a) What happens to steam which is produced in the early part of the fractional distillation?

(b) When a water and ethanol mixture is distilled the first distillate is not a pure liquid. Which liquid do you think forms the largest part of the first distillate?

15 You have to separate a mixture of three liquids by fractional distillation.
(a) How would you use the thermometer to help you?
(b) How many beakers would you need? Explain your answer.

3.4 Checking the purity of a substance

A pure substance melts at a particular temperature and boils at a particular temperature. If the substance has impurities it shows signs of melting over a range of temperatures, its melting point is not precise. The substance still has a definite boiling point but it is higher than that of the pure substance. Tables exist for the melting and boiling points of many pure substances. When a substance has been separated from a mixture its purity can be checked by finding the temperatures at which it melts and boils and comparing them with those in the tables.

The temperature of the substance is taken by immersing the bulb of the thermometer in the substance.

QUESTION

16 How does the way a substance melts and boils change when it contains impurities?

3.5 Summary

- A small range of chemical apparatus is used for a wide range of investigations (see p. 35).
- An element is a substance which cannot be broken down into simpler substances (see p. 38).
- The substances in a mixture can be separated by physical processes (see p. 38).
- The substances in a compound can be separated by chemical processes (see p. 38).
- Insoluble solids can be separated from liquids by decanting, filtering and centrifuging (see p. 39).
- A soluble solid can be separated from the solvent by evaporation or crystallisation (see pp. 41–2).
- A solvent can be separated from a solution and collected by distillation (see p. 42).
- Several soluble solids can be separated from a solution by chromatography (see p. 43).
- A few mixtures of two solids can be separated by sublimation (see p. 43).

- A separating funnel is used to separate two immiscible liquids (see p. 45).
- Fractional distillation is used to separate miscible liquids (see pp. 45–6).
- The purity of a substance can be checked by measuring its melting and boiling points (see p. 47).

◼ ☑ 4 Atomic structure

Objectives

When you have completed this chapter you should be able to:
- Define an **element**
- Identify the **symbols** for some widely used elements
- Describe the **structure of the atom** in terms of protons, neutrons and electrons
- Describe how the electrons are arranged in the first 20 elements in the **periodic table**
- Understand the terms **atomic number**, **mass number**, **isotope** and **relative atomic mass**.

4.1 The idea of elements

The alchemists worked with the idea that all matter was composed from four basic things called **elements**. These elements were fire, water, air and earth. Robert Boyle redefined the idea of the element in 1661 by saying that an element was any substance which could not be broken down into simpler substances by physical means or by chemical reactions. Following Boyle's definition of the element it is now known that there are 91 elements that occur naturally and another 21 have been made artificially.

(i) The element symbols

The names of many of the elements are quite long, so to save time in describing how the elements react and form compounds each element is given a symbol of either one or two letters. The first letter is always written as a capital letter. The letters in the symbol may relate to the elements' name in English – for example, the symbol for carbon is C. The letters may also relate to an older name for the element. For example, sodium was once known world wide as 'Natrium', and its symbol is Na.

Table 4.1 shows the symbols for the elements frequently featured in this book. See page 86 for the full list of element symbols displayed in the periodic table.

Element	symbol
hydrogen	H
helium	He
lithium	Li
carbon	C
nitrogen	N
oxygen	O
fluorine	Fl
neon	Ne
sodium	Na
magnesium	Mg
aluminium	Al
sulphur	S
phosphorus	P
silicon	Si
chlorine	Cl
potassium	K
calcium	Ca
iron	Fe
copper	Cu
zinc	Zn
bromine	Br
iodine	I
lead	Pb

Table 4.1 The element symbols

(ii) The idea of atoms

The idea that matter was made form tiny particles was first put forward by the Greek philosoper Democritus in about 400 BC. He reasoned that if you cut an object in half then cut one of the halves in half and so on you would eventually reach a particle of matter that was so small that it could not be divided. He called this particle an **atom** (atom means indivisible).

In 1808 John Dalton reinforced the idea of atoms when he found that he could explain his experimental results by saying that the chemicals taking part in the reactions existed as small particles.

At the beginning of the twentieth century experiments were performed to discover the structure of atoms. They continue to this day. Although over 200 sub atomic particles have been discovered the study of chemistry only requires a knowledge of three. They are the **proton, neutron** and **electron**.

When the ideas of elements and atoms are brought together it can be stated that each element has an atomic structure which is different from the atomic structures of all the other elements. It follows from this that the atoms of the different elements also vary in mass and in size.

QUESTION

I   How has the idea of elements changed from the days of alchemy?

(iii) The structure of the atom (Figure 4.1)

(a) The sub atomic particles

Dalton imagined an atom to be like a microscopic hard ball – like a snooker ball. When the structure of the atom was discovered it was found that an atom was certainly very small – about one ten millionth of a millemetre across – but it had two main parts – a central nucleus with a cloud of electrons moving around it. Further investigations showed that the nucleus is composed of one or more particles, called **protons**, and with the exception of most hydrogen atoms the nucleus has one or more particles, called **neutrons**. The protons and neutrons are known as **nucleons**.

The masses of the sub atomic particles are too small to be measured and expressed in grams but their masses can be compared. These comparisons are called relative masses and measured in atomic mass units (amu). A proton and a neutron have the same relative atomic mass but an electron is very much smaller with a relative atomic mass of about one two thousandth that of the nucleons (see Table 4.2).

Although a proton or a neutron is much more massive than an electron the space that they occupy at the centre of the atom is only very small when compared with the size of the atom. The volume of the nucleus is one thousand, million, million, million times smaller than the total volume of the atom. Most of the volume of the atom is due to the space occupied by the electrons which move around the nucleus at speeds approaching the speed of light.

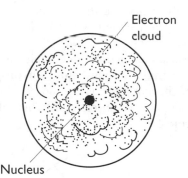

Figure 4.1 The structure of the atom

Particle	Relative mass (amu)	Relative charge
proton	1	1+
neutron	1	0
electron	0.0005 (1/2000)	1–

Table 4.2 Relative mass and electrical charges on sub atomic particles

If the nucleus was drawn to the same scale as the electron cloud in Figure 4.1 it would be too small to see with the naked eye.

(b) The charges on the particles

The electrons are held around the nucleus by **electrostatic forces**. Each electron has a negative electrical charge and each proton has a positive electrical charge. In the atoms of each element there is an equal number of protons and electrons. Their positive and negative charges balance. This makes the atom electrically neutral. Despite the electrostatic forces holding the electrons around the nucleus an atom can gain or lose electrons.

Table 4.2 summarises the relative mass and the electrical charges on the three main sub atomic particles.

QUESTIONS

    2 In what way is a proton and a neutron (a) similar, (b) different?

    3 In what ways are electrons and nucleons different?

4.2 A closer look at atomic structure

The electrons are grouped at different distances from the nucleus. This grouping is due to the amount or level of energy they possess. The region where each group of electrons occur is called a shell or energy level. The shell nearest the nucleus can hold up to two electrons, the second shell can hold up to eight electrons and the third electron shell, which is divided into two parts, can hold up to 18 electrons. The elements with the largest atoms have 7 electron shells.

The elements can be arranged in order of complexity of their atomic structure beginning with the one with the simplest atom – **hydrogen**.

Most hydrogen atoms have a nucleus composed of one proton and no neutrons. Around the nucleus moves an electron. This information can be represented as shown in Figure 4.2.

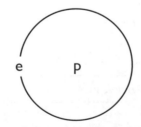

Figure 4.2   Nucleus and electron configuration of a hydrogen atom

The element which is next to hydrogen in complexity is **helium**. It has a nucleus with two protons and two neutrons. Two electrons move around the nucleus. This information can be represented as shown in Figure 4.3.

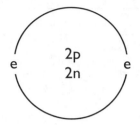

Figure 4.3   Nucleus and electron configuration of a helium atom

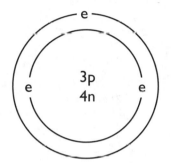

Figure 4.4   Nucleus and electron configuration of a lithium atom

The next element to helium is **lithium**. It has a nucleus with three protons and four neutrons. Three electrons move around the nucleus. This information can be represented as shown in Figure 4.4.

QUESTIONS

4   Beryllium is the next element to lithium in the order of complexity. It has four proton and five neutrons. (a) How many electrons will it have? Explain your answer. (b) Draw a diagram of a beryllium atom in the style of Figures 4.3–4.4.

5   When you have read the sections on atomic number, mass number and examined Table 4.3, draw diagrams for carbon, sodium and chlorine.

The arrangement of the electrons in the shell can also be displayed simply as a row of numbers. Table 4.3 features the first 20 elements of the periodic table with the numbers of electrons in each of their shells.

4.3 Atomic number

The atoms of an element all have the same number of protons. This number is called the **atomic number** of the element. For example the atoms of hydrogen have one proton so the atomic number of hydrogen is 1. The atomic number can be written as a subscript below the symbol for the element as

$_1$H

Element	Electrons in their shells			
hydrogen	1			
helium	2			
lithium	2	1		
beryllium	2	2		
boron	2	3		
carbon	2	4		
nitrogen	2	5		
oxygen	2	6		
fluorine	2	7		
neon	2	8		
sodium	2	8	1	
magnesium	2	8	2	
aluminium	2	8	3	
silicon	2	8	4	
phosphorus	2	8	5	
sulphur	2	8	6	
chlorine	2	8	7	
argon	2	8	8	
potassium	2	8	8	1
calcium	2	8	8	2

Table 4.3 Electrons and shells

The atomic number may also be known at the proton number. It has its own symbol. This is Z.

QUESTION

6 Helium has two protons, nitrogen has seven and potassium has 19. Display their atomic numbers with the symbols as shown for hydrogen. (Look at Table 4.1 (p. 50) for the symbols of the elements.)

4.4 The mass number

This number is the sum of the protons and neutrons in an atom of an element. For example the sodium atom has 11 protons and 12 neutrons and so has a mass number of $11 + 12 = 23$. The mass number is written as a superscript above the symbol for the element as

^{23}Na

The mass number is also known as the **nucleon number**. It has the symbol A.

4.5 Finding the neutrons in an atom

If the atomic number Z and the mass number A of an atom are known the number of neutrons can be found by the following subtraction:

number of neutrons $= A - Z$

7 What is the number of neutrons in an atom of (a) magnesium $^{24}_{12}Mg$ (b) copper $^{64}_{29}Cu$

4.6 Isotopes

The number of protons in the atoms of an element are fixed. All atoms of hydrogen, for example, have one proton and therefore the element has an atomic number of 1. The number of neutrons in the atom of an element is not fixed. For example, although most atoms of hydrogen do not have a neutron some have one neutron and a few have two. These atoms of an element which have different numbers of neutrons are called **isotopes**. They all have the same atomic number but have different mass numbers, as the isotopes of hydrogen show:

hydrogen   deuterium   tritium
^1_1H ^2_1H ^3_1H

QUESTIONS

8 Do the isotopes of an element have the same (a) atomic number, (b) mass number? Explain your answer.
9 The atomic number of magnesium is 12 and it has isotopes with 12, 13 or 14 neutrons. Write down the three isotopes of magnesium showing their atomic number and mass number.

4.7 Radioisotopes

The isotopes of some elements are not stable. This means that they break down or decay to produce a daughter isotope which is an element that is different from the original element. When an unstable isotope decays one or more types of radiation are released. There are three types of radiation. They are:

(i) Alpha particle radiation

This is formed by a particle made from two protons and two neutrons. This kind of radiation is usually produced by elements which have a large nucleus. They have a relative atomic mass which is greater than 210. When an element loses an alpha particle its atomic number falls by 2 and its mass number falls by 4. For example when uranium 238 ($^{238}_{92}U$) loses an alpha particle it forms thorium 234 ($^{234}_{90}Th$)
An alpha particle has two positive electrical charges and travels at 11 904 km/sec. It is stopped by 3 cm of air or a few sheets of paper.

(ii) Beta particle radiation

This occurs usually in the unstable nuclei of elements with a relative atomic mass below 210. It is produced when a neutron in a nucleus changes into a proton. The atomic number of an atom increases by 1 when it releases a beta particle.

A beta particle is the size of an electron and has one negative charge. It travels at a speed of 148 800 km/sec and can be stopped by a few millimetres of plastic or metal. It can also penetrate a few centimetres of living tissue.

(iii) Gamma radiation

This is usually produced when alpha and beta particles are released. It is a wave of electromagnetic energy with a wave length that is shorter than those of X-rays. It does not change the atomic number or mass number of the atom. Gamma radiation travels at about 300 000 km/sec – light speed – and can pass through the whole bodies of living things. It can be stopped by 5–25 cm of lead or up to to 3 m of concrete.

QUESTIONS

10 Which forms of radiation result in a change in structure in the nucleus of an atom?

11 Which form of radiation results in a loss of energy from the atom?

12 Carbon 14 decays to form nitrogen. Look at Table 4.3 (p. 54) to explain how this change takes place

4.8 Uses of radioisotopes

(i) In industry

The beta and gamma radiation from an isotope can be formed into a beam which is directed at a detector. When a material is placed in the beam the amount of radiation reaching the detector in a certain time falls. This change in radiation reaching the detector is related to the thickness of the material. The radioactive isotope can be used to check the thickness of a material that is being produced in sheet form such as metal, paper or plastic. If the sheet-making machine begins to produce sheets that are too thick or too thin this is recorded by the detector which then sends the information to the part of the machine that controls the rollers. The information is used by the machine to reset the rollers automatically so that the correct thickness of the material is produced again.

(ii) In medicine

As beta and gamma radiation can pass into living tissue and destroy cells they can be used to kill cancer cells. The radioisotope strontium 90 produces beta radiation which is used to treat skin cancers. The radioisotope cobalt 60 produces gamma radiation which is used to destroy cancers which have formed deep inside the body.

QUESTIONS

13 How do you think a beam of radiation could be used to tell if bottles are filled to the required level on a production line?

14 Why is gamma radiation used instead of beta radiation for destroying cancers deep inside the body?

4.9 The relative atomic mass

The mass of an atom is very, very small. A one millilitre drop of water has a mass of one gram but one atom of hydrogen in a water molecule has a mass of 0.000 000 000 000 000 000 000 002 g. This figure is too difficult to work with so the relative atomic mass, RAM or Ar, has been devised to make calculations easier. When the relative atomic mass was first devised, hydrogen was used as the element by which all others were compared. Today the most common isotope of carbon, called carbon 12, is used. An atom of carbon 12 has six protons and six neutrons. They give it an atomic mass of 12. When the mass of carbon 12 and hydrogen are compared, hydrogen has a mass 1/12 that of carbon 12. This mass is still used to compare with the masses of all the other elements. It is stated in the equation for finding Ar of an element:

$$\text{Ar of an element} = \frac{\text{Mass of one atom of the element}}{1/12 \text{ mass of one atom of carbon 12}}$$

The equation can be rearranged to:

Mass of one atom of the element

= Ar of the element × 1/12 mass of one atom of carbon 12

Using the equation and Table 4.4, it can be seen that:

Element	Symbol	Ar
hydrogen	H	1
carbon	C	12
nitrogen	N	14
oxygen	O	16
sodium	Na	23
magnesium	Mg	24
sulphur	S	32
chlorine	Cl	35.3
copper	Cu	63.5

Table 4.4 Ar of some elements

- The mass of one atom of hydrogen is $1 \times 1/12 = 1/12$ the mass of a carbon 12 atom or that the mass of a carbon 12 atom is 12 times the mass of a hydrogen atom.
- The mass of an oxygen atom is $16 \times 1/12 = 1.33$ times greater than the mass of a carbon 12 atom.
- The mass of a magnesium atom is $24 \times 1/12 = 2$ times greater than a carbon 12 atom.

Samples of atoms of an element may contain isotopes. The proportion of the different isotopes in a typical sample of the element must be worked out and used to calculate the RAM of the element. The RAM of an element is calculated by multiplying the percentage of each isotope with its RAM, then adding them together and dividing by one hundred.

EXAMPLE

In a typical sample of chlorine 75% of the atoms are of isotope chlorine 35 and 25% are of isotope chlorine 37. In a typical sample of chlorine the total value of the RAM is

$$(35 \times 75) + (37 \times 25) = 3550 \text{ units}$$

The value of the RAM of one atom of chlorine in this sample is 3550/100 = 35.5

QUESTION

15　Why is the RAM of some elements not a whole number?

4.10 Summary

- An element is a substance which cannot be broken down into simpler substances (see p. 49).
- Each element has a symbol of either one or two letters (see p. 49).
- The atom has sub atomic particles called protons, neutrons and electrons (see p. 50).
- The number of protons in an atom is the atomic number of the atom (see p. 53).
- The mass number of an atom is the sum of the protons and the neutrons in the atom (p. 54).
- The atoms of an element which have different numbers of neutrons are called isotopes (see p. 55).
- Some isotopes produce radiation (see p. 55).
- The relative atomic mass is used to compare the masses of atoms (see p. 57).

■ ⌄ 5 Bonding

Objectives

When you have completed this chapter you should be able to.
- Understand diagrams showing the **electronic structure of atoms**
- Describe the **ionic bond**
- Describe **giant ionic structures** and some of their physical properties
- Describe the **covalent bond**
- Understand the concept of the **molecule**
- Describe some physical properties of **simple molecular structures**
- Describe **giant covalent structures** and some of their physical properties.
- Understand the structure and some of the physical properties of **metals**
- Understand the structure and some of the properties of **plastics**.

5.1 When atoms combine

Atoms of elements rarely exist on their own. They join together either with each other or with atoms of other elements. When atoms of different elements join together they form bonds between themselves and make new substances called **chemical compounds**. These substances are not mixtures and the elements in them cannot be separated by filtering or distillation but they can be separated by chemical reactions.

5.2 Representing electronic structures

The electronic structure of an atom can be represented by a diagram. Figure 5.1 shows the electron structure of a lithium atom.

Each circle represents an electron shell or energy level. When the electrons in two atoms are being considered together, dots are used for the electrons in one atom and crosses are used for electrons in the other atom. In diagrams where an atom is large or only the changes in the outer shells are being considered the diagrams may show just the two outer shells of electrons.

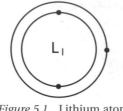

Figure 5.1 Lithium atom

5.3 Electrons in the outer shell

An element with atoms that have an outer shell full of electrons is very stable and unreactive. These elements are known as the **noble gases**. They are also known as the **inert gases** because they take part in very few chemical reactions. All the other elements do not have full outer shells and are much more reactive.

When a chemical reaction takes place between these elements, the atoms taking part fill their outer shells and become stable like inert gases. They may attain a full outer shell by either gaining one or more electrons from another atom, by sharing electrons with another atom, or by releasing one or more electrons to another atom so the full shell beneath becomes the outer shell (Figure 5.2).

5.4 The ionic bond

An ionic bond forms when one or more electrons move between atoms so that their outer shells become full. Common salt – sodium chloride – is held together by ionic bonds. They form in the following way:

(i) Making full shells

A sodium atom has an outer shell with only one electron in it. If it loses this electron its full lower shell becomes the outer shell and the atom has the stability of an atom of the inert gas neon (Figure 5.3).

The chlorine atom has room for one more electron in its outer shell and readily accepts an electron from a sodium atom. The outer shell of the chlorine atom becomes full and the atom has the stability of an atom of the inert gas called argon (Figure 5.4).

(ii) Generating charge

The loss of the electron from the sodium atom gives the atom a positive charge because the nucleus has 11 positively charged protons surrounded by only 10 negatively charged electrons.

The transfer of the electron to the chlorine atom gives the atom a negative

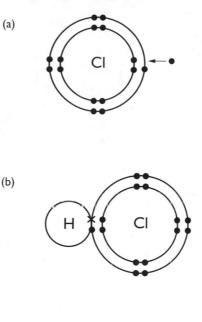

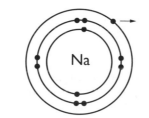

(c)

Figure 5.2 (a) An atom gaining an electron (b) Atoms sharing electrons (c) Atom releasing one electron to reveal a full shell beneath

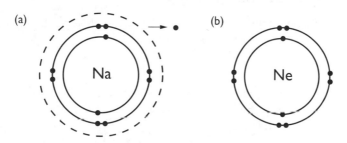

Figure 5.3 (a) Electron model of a sodium atom losing an electron (b) Electron model of neon for comparison

charge because the nucleus has 17 positively charged protons surrounded by 18 positively charged electrons.

The electrically charged atoms are called ions. An ion with a positive charge like sodium is called a cation and an ion with a negative charge like a chloride ion is called an anion.

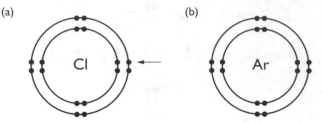

Figure 5.4 (a) Electron model of a chlorine atom losing an electron (b) Electron model of argon for comparison

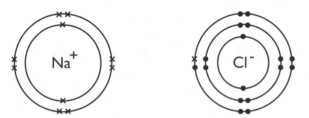

Figure 5.5 Sodium ion and chloride ion close to each other

An ion is shown by placing positive or negative signs after the symbol. For example:

sodium ion Na^+ chloride ion Cl^-

The electronic structure of a sodium ion and a chloride ion can also be represented as a diagram or as a number sequence (Figure 5.5).

The number sequence for the sodium ion is 2, 8. and the number sequence for the chloride ion is 2, 8, 8.

(iii) Forming the bond (Figure 5.6)

The force of attraction between the positively and negatively charged ions forms the ionic bond which binds them together.

QUESTION

1 (a) What is the number sequence for the electron structure of the sodium atom and the chlorine atom? (b) Why are these different from the ions? (Check your answer with Table 4. 3.)

(iv) Two more examples of ionic bonding

(a) Magnesium oxide

When magnesium burns in air (see Figure 5.6) each magnesium atom loses two electrons to an oxygen atom. The magnesium ion has two positive charges and the oxygen ion has two negative charges. The force of attraction between them binds the ions together to form magnesium oxide – a white powder.

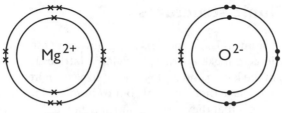

Figure 5.6   Ions of magnesium oxide

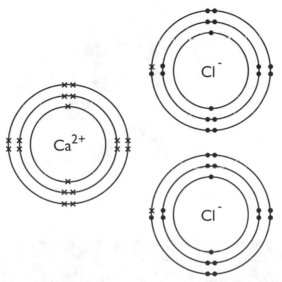

Figure 5.7   Ions of calcium chloride

The sequence of the electrons in the two ions are magnesium ion 2, 8. and oxygen ion 2, 8.

QUESTION

2   (a) How has the atomic structure of the magnesium and oxygen atoms changed when they form the compound magnesium oxide? (b) Which element does the electronic configurations resemble, and why have the atoms changed in this way?

(b) Calcium chloride

When calcium chloride is formed the calcium atoms lose two electrons from their outer shells. Each electron joins the electrons in the outer shell of a chlorine atom to make a chloride ion. Two chloride ions are attracted to each calcium ion and bond to form calcium chloride which is used as a drying agent in science laboratories (Figure 5.7).

The sequence of electrons in the two ions are calcium ion 2, 8, 8 and chloride ions 2, 8, 8.

5.5 Giant ionic structures

The billions of sodium and chloride ions in a grain of salt are arranged alternately in three dimensions to make a structure called a **lattice**. There is no maximum number of ions in the lattice and as it is a huge size compared with a single atom it is known as a giant structure. The ionic bonds bind the ions together so closely in such huge numbers that they form a solid substance. The arrangement of the ions in the lattice give the substance its crystalline shape (Figures 5.8, 5.9).

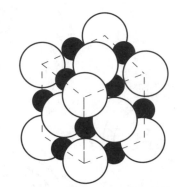

Figure 5.8   Picture of a lattice

Figure 5.9   Salt crystals

5.6 The physical properties of giant ionic structures

The ionic bonds hold the ions together so strongly that a great deal of heat energy is needed to separate the ions and turn the substance from a solid into a liquid. Substances made from an ionic lattice have high melting points and high boiling points. Sodium chloride, for example, melts at 808 C.

The electrons are so tightly bound to the ions that they are not free to move through the substance and it does not conduct electricity.

If the substance is dissolved in water or is melted it can conduct electricity because its ions split up and can move freely.

QUESTION

3 Why can the molten or dissolved ionic substance conduct electricity but the solid form cannot?

5.7 The covalent bond

This bond forms when two atoms share one or more electrons in their outer shell to attain the stability of an inert gas.

A hydrogen atom, for example, has only one electron. The atoms join together in pairs and share their electrons so that each one has two electrons in its shell. This gives each atom the stability of the inert gas called helium (Figure 5.10).

The sharing of the electrons generates a strong force of attraction between the two atoms which forms the covalent bond that holds the atoms together. Atoms which are held together by covalent bonds form a group called a **molecule**. A pair of electrons being shared between two atoms is one covalent bond.

Covalent bonds are usually represented by a short dash (-) between the symbols of atoms they link together. The covalent bond in the hydrogen molecule is shown in Figure 5.11.

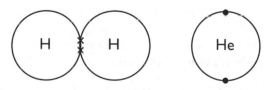

Figure 5.10   A hydrogen molecule and a helium atom

H—H

Figure 5.11   The covalent bond in a hydrogen molecule

5.8 Examples of small molecules

(i) Hydrogen chloride

A molecule of hydrogen chloride is composed of one atom of hydrogen and one atom of chlorine (Figure 5.12).

The covalent bond in the hydrogen chloride molecule is shown in Figure 5.13. Hydrogen chloride is a strong-smelling, colourless gas which fumes in moist air.

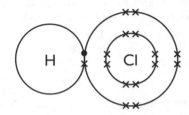

Figure 5.12   A hydrogen chloride molecule

$$H—Cl$$

Figure 5.13   The covalent bond in the hydrogen chloride molecule

(ii) Water

A molecule of water is composed of two atoms of hydrogen and one atom of oxygen (Figure 5.14).

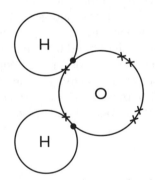

Figure 5.14   A water molecule

The covalent bonds in the water molecule are shown in Figure 5.15. Pure water is a colourless liquid without a smell.

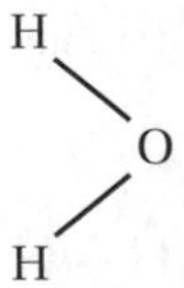

Figure 5.15   The covalent bonds in the water molecule

(iii) Ammonia

A molecule of ammonia is composed of three atoms of hydrogen and one atom of nitrogen (Figure 5.16).

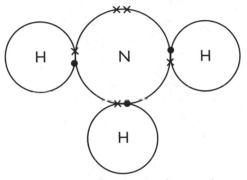

Figure 5.16 A molecule of ammonia

Figure 5.17 The covalent bonds in the ammonia molecule

The covalent bonds in a molecule of ammonia are shown in Figure 5.17. Ammonia is a strong-smelling, colourless gas.

(iv) Methane

A molecule of methane is composed of four atoms of hydrogen and one atom of carbon (Figure 5.18).

The covalent bonds in a methane molecule are shown in Figure 5.19.

Methane is a colourless gas without a smell which forms an explosive mixture with air.

(v) Oxygen

A molecule of oxygen is made up of two oxygen atoms which each share two electrons with the other atom to give it the stability of the inert gas neon (Figure 5.20).

The covalent bonds in an oxygen molecule is shown in Figure 5.21.

Oxygen is a colourless gas without a smell.

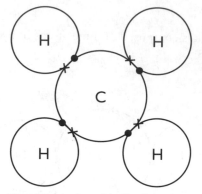

Figure 5.18   A molecule of methane

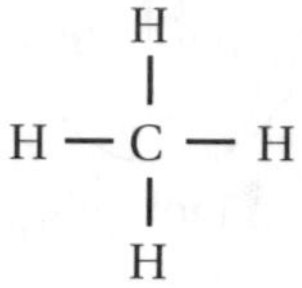

Figure 5.19   The covalent bonds in the methane molecule

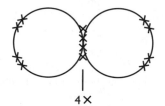

4✗

Figure 5.20   A molecule of oxygen

$$O=O$$

Figure 5.21   The covalent bonds in the oxygen molecule

QUESTIONS

4   What happens to the electron structure of (a) a hydrogen atom, and (b) an oxygen atom when a molecule of water forms?
5   Tetrachloromethane is a molecule formed from one atom of carbon and four atoms of chlorine. Draw a diagram of this molecule showing the electron structure of the atoms.

5.9 The physical properties of simple molecular structures

Most of the compounds formed by covalent bonds have small molecules containing a small number of atoms. These molecules are held together by intermolecular forces which are much weaker than ionic or covalent bonds. Much less heat energy is needed to separate the molecules compared to the heat energy required to separate the atoms and the substances made from these small molecules have low melting points and boiling points. Oxygen for example melts at −219°C and boils at − 183°C.

5.10 Giant covalent structures

A few substances have atoms which are held in giant structures by covalent bonds. These substances form solids.

Diamond is formed from carbon atoms which each share four electrons to make a lattice (Figure 5.22).

Silicon atoms form covalent bonds with oxygen atoms to form silica which has a more complicated lattice than diamond (Figure 5.23).

Graphite is formed from carbon atoms which only share three of their electrons. Their fourth electrons spread out over the whole structure (Figure 5.24).

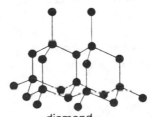

diamond

Figure 5.22   A diamond lattice

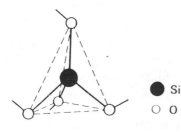

● Si

○ O

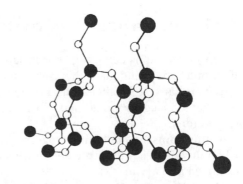

Figure 5.23   Part of a silica lattice with some atoms removed to make the bonding clearer

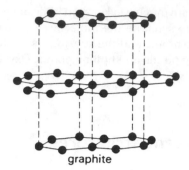
graphite

Figure 5.24 The structure of graphite

5.11 The physical properties of giant covalent structures

The covalent bonds hold the atoms together so strongly that a great deal of energy is needed to separate them and they have high melting points and boiling points. Diamond, for example, melts at 3550°C.

The lattice structure of diamond and silica, in which every available electron forms a bond, makes these substances very tough and hard. The lack of free electrons make these substances unable to conduct electricity.

In **graphite**, where only three out of a possible four electrons form bonds, the atoms are arranged in layers which slide over each other and make a soft substance. The free electrons which are not used in bonding allow graphite to conduct electricity.

QUESTION

6 How does the presence of ionic and covalent bonds in a substance affect its melting point and boiling point? Explain your answer.

5.12 The structure of metals

When metal atoms come together in large numbers they lose the electrons in their outer shells and make themselves more stable. The electrons are not transferred or shared with other atoms but flow freely round the atoms and bind them together to make the metal solid. The atoms are often described as being in a 'sea of electrons'.

As the atoms are not bonded together they can slip and slide past each other when the metal is struck, pressed or pulled. This allows the metal to be shaped or drawn into a wire.

The electrons moving between the atoms can carry electricity or heat through the metal and make it an excellent conductor (Figures 5.25 and 5.26).

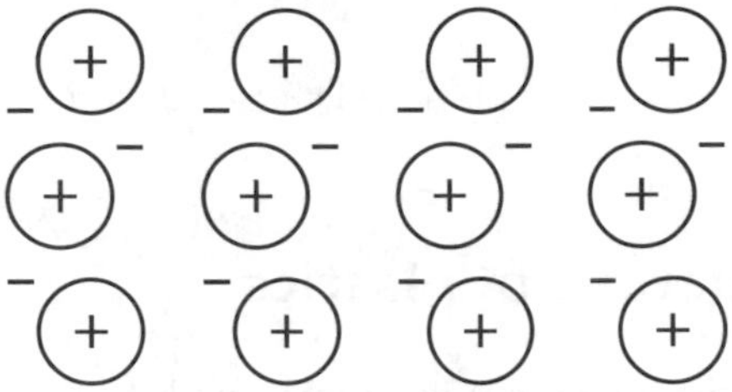

Figure 5.25   Metal atoms packed together with free electrons moving round them

Figure 5.26   An intricate metal design due to ability of metal to be shaped

7 Why can (a) graphite, (b) metals conduct electricity?

5.13 The structure of plastics

Plastics are made from **polymers**. A polymer is made from atoms which are joined together by covalent bonds to form long chain molecules (Figure 5.27).

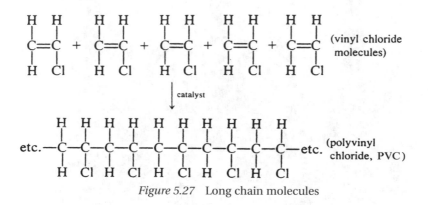

Figure 5.27   Long chain molecules

(i) Thermosoftening plastics or thermoplastics

In polythene, the molecules are tangled round each other and are only held together by weak inter-molecular forces. When polythene is warmed the forces are weakened even more and the molecules can move away from each other. This movement of the molecules allows the shape of the warm material to be changed. When the polythene cools, the molecules are held together again by the inter-molecular forces until the plastic is heated once more.

(ii) Thermosetting plastics

Melamine is a plastic that sets hard after it has been heated. This happens because the covalent bonds form between the molecules and hold them firmly in place. The bonds are not broken if the the plastic is heated again and the plastic does not melt.

Plastics are also considered in the context of industrial materials (see polymers structures and uses, p. 233).

8   What happens when polythene and melamine are warmed? Explain your answer.

5.14 Summary

- The electronic structure of an atom can be represented by a diagram (see p. 59).
- An ionic bond forms when one or more electrons move between atoms so that their outer shells become full (see p. 60).
- Giant ionic structures can be formed which have high melting and boiling points (see p. 64).
- A covalent bond forms when two atoms share one or more electrons (see p. 65).
- Simple molecular structures have low melting and boiling points (see p. 68).
- Giant covalent structures have high melting and boiling points (see p. 69).
- A 'sea of electrons' surrounds atoms in a piece of metal (see p. 70).
- Plastic materials are made from long chain molecules (see p. 72).

■ 🗹 6 Representing reactions

Objectives

When you have completed this chapter you should be able to:
- Understand a **word equation**
- Construct formulae of some **ionic compounds**
- Construct formulae of some **covalent compounds**
- Understand a **symbol equation**
- **Balance** simple equations
- **Write and balance** simple equations
- Understand **ionic equations**
- Understand **half equations**.

6.1 Chemical reactions

There are many types of chemical reactions. For example, when two or more compounds react to form one compound an **addition reaction** is said to have taken place. When a compound is heated and breaks down to form other compounds a **thermal decomposition** is said to have taken place.

The elements and compounds which take part in the reaction are called the reactants and the substances that are formed are called the **products**. This can be more simply written as:

Reactants form Products

and shown even more simply as:

Reactants → Products

6.2 The word equation

The names of the reactants and products can then be written on each side of the arrow to make a word equation for the reaction.

Where there are two or more names of substances on one side of the arrow they are linked together by a plus (+) sign.

For example, the chemical reaction which takes place between carbon and oxygen to produce carbon dioxide is written as:

carbon + oxygen → carbon dioxide

6.3 The reversible reaction

Some reactions are **reversible**. This means that when the reactants have produced the products, the products can then produce the reactants. An equation for a reversible reaction has a double arrow ($\rightleftharpoons$).

The reversible reaction is divided into two parts. The first part is the **forward reaction**, which is the reaction written and read from left to right. The second part is the **back reaction**, which is revealed by reading the equation from right to left. For example, when carbon dioxide dissolves in water some of it reacts with the water and forms carbonic acid:

carbon dioxide + water ($\rightleftharpoons$) carbonic acid:

QUESTIONS

   1 Write down the word equations for the following chemical reactions:
      (a) Magnesium burns in oxygen to form magnesium oxide
      (b) Sodium hydrogen carbonate breaks down to sodium carbonate, water and carbon dioxide when it is heated
      (c) Barium chloride and sodium sulphate solutions form barium sulphate and sodium chloride when they are mixed together.
   2 What is the back reaction of the reversible reaction featuring carbon dioxide, water and carbonic acid?

6.4 The symbol equation

The word equation can be replaced by the symbol equation.

This gives more information about the reactants and the products and is much quicker to write once you become familiar with the formulae of compounds and how to balance a symbol equation.

The words to describe a substance can be replaced by a formula which uses the symbols for the elements and shows the number of atoms taking part in the reaction. For example the word equation:

carbon + oxygen $\rightarrow$ carbon dioxide

can be written as:

$$C + O_2 \rightarrow CO_2$$

The following sections show how you can work out the formulae of some simple compounds and use them to make symbol equations.

6.5 Finding the formula of a compound

(i) Ionic compounds

To write the correct formula for an ionic compound the charges on the ions taking part need to be studied to produce a compound that is electrically neutral.

EXAMPLE

When sodium burns in chlorine gas sodium chloride is formed

The charge on a   The charge on a
sodium ion is 1^+ chlorine ion is 1^-

So 1 sodium ion   joins with   1 chloride ion

to give the formula for sodium chloride = Na Cl

When aluminium sulphide is formed

The charge on an   The charge on a
aluminium ion is 3^+ sulphide ion is 2^-

So 2 aluminium ions join with 3 sulphide ions
to give a formula for aluminium sulphide = Al_2S_3

The way of calculating the ionic formula is

A ion +charge number   B ion −charge number

−charge number A ion   +charge number B ion

When the positive and negative charges are the same

e.g. $Zn^{2+} SO_4^{2-}$

the numbers cancel out and the formula of the ionic compound is simply
the symbols and formulae joined together $ZnSO_4$.

QUESTION

3   Use Table 6.1 to answer these questions. What is the formula of (a) silver nitrate,
    (b) calcium hydroxide, (c) zinc phosphate, (d) aluminium hydroxide, (e) silver
    chloride, (f) calcium sulphate, (g) iron (II) hydroxide?
    Note: Ions of two or more elements are enclosed in a bracket when two or
    more of the ions are present in a compound, For example $(NH_4)_2SO_4$ and
    $Ca_3(PO_4)_2$

Charges on positive ion			Charges on negative ion		
Na^{1+}	Ca^{2+}	Al^{3+}	Cl^{1-}	SO_4^{2-}	PO_4^{3-}
sodium	calcium	aluminium	chloride	sulphate	phosphate
Ag	Zn		Br	O	
silver	zinc		bromide	oxide	
NH_4	Fe	Fe	NO_3		
ammonium	iron (II)	Iron (III)	nitrate		
Cu	Cu		OH		
copper (I)	copper (II)		hydroxide		

Table 6.1   Charges on common ions

(ii) Covalent compounds

Table 6.2 shows the number of electrons shared by some non-metals. It should be noted that sulphur can also share four or six electrons and that phosphorus can also share five electrons.

Hydrogen shares one electron, so using this information and the information in Table 6.2 the formula for water and methane can be worked out as shown in the following examples.

The system for working out the formulae of covalent compounds is similar to that used for working out the formulae for ionic compounds as these examples show:

EXAMPLE 1: WATER

      hydrogen atoms share one   oxygen atoms share
      electron   two electrons

      so two hydrogen atoms share with one oxygen atom
      to give a formula for water of H_2O

EXAMPLE 2: METHANE

      carbon atoms share four   hydrogen atom share
      electrons   one electron

      so one carbon atom shared with four hydrogen atoms
      to give a formula for methane of CH_4

QUESTION

    4 Use Table 6. 2 to answer these questions. What is the formula for a compound between (a) silicon and oxygen, (b) phosphorus (as shown in the table) and chlorine?

6.6 The ratio of atoms in a formula

In both ionic and covalent compounds the numbers in the formula give the ratio of the atoms of the different elements.

For example, the formula CO_2 shows that there are twice as many oxygen atoms than carbon atoms in a molecule of carbon dioxide.

No. of electron shared			
1	2	3	4
F fluorine	O oxygen	N nitrogen	C carbon
Cl chlorine	S sulphur	P phosphorus	Si silicon

Table 6.2 Number of electrons shared

QUESTION

 5 What elements are present in each of these compounds and what is the ratio
   between the elements? Use Table 4.1 on p. 50 to identify the elements (a) $CaCl_2$,
   (b) HCl, (c) NH_3, (d) SO_2, (e) CH_4, (f) C_2H_6.

6.7 Using chemical formulae

A more detailed description of a chemical reaction revealed by the symbol equation shows the ratio of the atoms in a molecule of a compound. In the following
equation it can be seen that carbon dioxide has two atoms of oxygen for one atom
of carbon:

   $C + O_2 \rightarrow CO_2$

6.8 State symbols

A symbol equation is completed by adding state symbols to each of the reactants
and products.
   The state symbols are –(s) = solid, (l) = liquid, (g) = gas, (aq) = aqueous solution.
They are placed in the equation after each reactant and product.

EXAMPLE

   In the reaction between carbon and oxygen

      carbon + oxygen → carbon dioxide

   the state symbols are

      $C (s) + O_2 (g) \rightarrow CO_2 (g)$

   and in the reaction between sodium chloride and sulphuric acid

      sodium chloride + sulphuric acid → sodium hydrogen sulphate
      + hydrogen chloride

   the state symbols are

      $NaCl (s) + H_2SO_4 (aq) \rightarrow NaHSO_4 (aq) + HCl (g)$

QUESTIONS

 6 What are the reactants and products taking part in the chemical reactions 1,2
   and 3? Use Table 6.3 to identify the substances:
   (1) $Na (s) + H_2 (l) \rightarrow NaOH (aq) + H_2 (g)$
   (2) $CaCO_3 (s) \rightarrow CaO (s) + CO_2 (g)$
   (3) $NaOH (aq) + HNO_3 (aq) \rightarrow NaNO_3 (aq) + H_2O (l)$
 7 In what state is sodium in equation (1)?
 8 In what state is hydrogen in equation (1)?
 9 In what state is calcium oxide in equation (2)?
 10 In what states are sodium hydroxide and water in equation (3)?

Formula	Substance
H_2	hydrogen
CO_2	carbon dioxide
HNO_3	nitric acid
CaO	calcium oxide
H_2O	water
NaOH	sodium hydroxide
$CaCO_3$	calcium carbonate
$NaNO_3$	sodium nitrate
Na	sodium

Table 6.3   Reactants and products

6.9 The balanced equation

In a chemical reaction the elements may recombine but they are not destroyed. There are the same number of atoms of each element at the end of the reaction as there was at the beginning. This means that in the symbol equation there must be the same number of atoms of each element on both sides of the arrow.

EXAMPLE

$$ C \qquad + O_2 \rightarrow \qquad CO_2 $$

one atom $\qquad$ two atoms $\qquad$ one atom two atoms
of carbon $\qquad$ of oxygen $\qquad$ of carbon of oxygen

Most equations do not balance as easily as this one when the formulae are written down.

They are balanced in the following way:

(1) Each formula is checked that it is written down correctly.
(2) Each element is checked to find if the same number of atoms are present on each side of the equation.
(3) When a difference is found it is corrected by multiplying the formulae in the equation. It can never be corrected by changing the numbers in a formula.

EXAMPLE

When the word equation

hydrogen + oxygen → water

is replaced by a symbol equation

$$ H_2 \text{ (g)} + O_2 \text{ (g)} \rightarrow H_2O \text{ (l)} $$

the number of hydrogen atoms on both side of the equation balance but the number of oxygen atoms do not.

By multiplying the formula for water by 2 the number of oxygen atoms on both sides of the equation balance:

$$H_2 \text{ (g)} + O_2 \text{ (g)} \rightarrow 2H_2O \text{ (l)}$$

but the number of hydrogen atoms do not.

The number of hydrogen atoms is balanced by multiplying the formula for hydrogen by 2:

$$2H_2 \text{ (g)} + O_2 \text{ (g)} \rightarrow 2H_2O \text{ (l)}$$

Further checking shows that the equation is balanced for both hydrogen and oxygen atoms.

QUESTIONS

11 Balance each of these equations:
 (a) $Cu \text{ (s)} + O_2 \text{ (g)} \rightarrow CuO \text{ (s)}$
 (b) $H_2 \text{ (g)} + Cl_2 \text{ (g)} \rightarrow HCl \text{ (g)}$
 (c) $NaHCO_3 \text{ (s)} \rightarrow Na_2CO_3 \text{ (s)} + H_2O \text{ (l)} + CO_2 \text{(g)}$
 (d) $Ag_2O \text{(s)} + H_2O_2 \text{ (aq)} \rightarrow Ag \text{ (s)} + H_2O \text{ (l)} + O_2 \text{ (g)}$
 (e) $Al \text{ (s)} + Cl \text{ (g)} \rightarrow Al Cl_3 \text{ (s)}$
12 Write balanced symbol equations from these word equations (use Table 6.4 to help you)
 (a) Potassium + water $\rightarrow$ potassium hydroxide + hydrogen.
 (b) Magnesium carbonate + hydrochloric acid $\rightarrow$ magnesium chloride + water + carbon dioxide
 (c) Iron (III) oxide + carbon monoxide $\rightarrow$ iron + carbon dioxide

6.10 Ionic equations

A reaction may also be examined by considering the ions taking part. For example, when hydrochloric acid and sodium hydroxide react together a neutralisation reaction takes place and water is produced.

Substance	Formula/symbol	State symbol
hydrogen	H_2	g
water	H_2O	l
potassium	K	s
potassium hydroxide	KOH	aq
magnesium carbonate	$MgCO_3$	s
magnesium chloride	$MgCl_2$	s
carbon dioxide	CO_2	g
hydrochloric acid	HCl	aq
iron (III) oxide	Fe_2O_3	s
iron	Fe	s
carbon monoxide	CO	g

Table 6.4 The symbol, formulae and state symbols of a range of substances

This is represented by:

(1) The word equation

hydrochloric acid + sodium hydroxide → sodium chloride + water

(2) The symbol equation

HCl (aq) + NaOH (aq) → NaCl (aq) + H_2O (l)

(3) The ions in the equation

H^+ (aq) + Cl^- (aq) + Na^+ (aq) + OH^- (aq) → Na^+ (aq) + Cl^- (aq) + H_2O (l)

It can be seen from equation (3) that two ions are changed in the reaction. They are the hydrogen ion (H^+) and the hydroxide ion (OH^-). It can also be seen that two ions are not changed in the reaction. They are the sodium ion (Na^+) and the chloride ion (Cl^-). These ions which remain unchanged during a chemical reaction are called **spectator ions**. They can be removed from the equation to produce an equation which features only the ions which are changed during the reaction. For example when the spectator ions are removed from equation (3) the following equation is produced:

(4) H^+ (aq) + OH^- (aq) → H_2O (l)

This equation which does not feature the spectator ions is called an **ionic equation**.

Ionic equations must balance by having the same number of atoms and the same number of charges on both sides of the equation. In equation (4) the two opposite charges shown on the left have cancelled each other out on the right when the water molecule formed.

QUESTION

13 Equation (1) describes the reaction between iron (II) sulphate and sodium hydroxide.
(1) $FeSO_4$ (aq) + $2NaOH$ (aq) › $Fe(OH)_2$ (s) + Na_2SO_4 (aq)
The ions of the compounds in this reaction are
(2) Fe^{2+} (aq) + SO_4^{2-} (aq) + $2Na^+$ + $2OH^-$ (aq) → Fe^{2+} $(OH^-)^2$ (s) + Na^{2+} (aq) SO_4^{2-} (aq)
(a) Which ions in the reaction are spectator ions?
(b) Construct an ionic equation for this reaction.

6.11 Electrolysis and half equations

In a solid ionic substance the ions are bonded together and cannot move or conduct electricity. In a liquid ionic substance, which is made by melting the substance or dissolving it in water, the ions are free to move and can conduct an electric current through the liquid. A liquid which contains ions and conducts electricity is called an electrolyte (Figure 6.1).

When the switch is closed in the circuit electrons flow from the power source to the cathode (negative electrode). The positive charged ions in the electrolyte

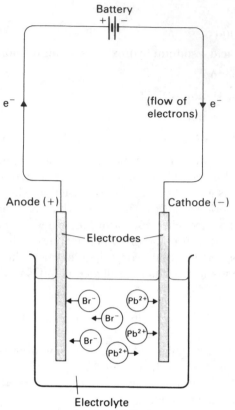

Battery

Anode (+)

Cathode (−)

e^- (flow of electrons) e^-

Electrodes

Br⁻   Pb²⁺
Br⁻
Br⁻   Pb²⁺
Pb²⁺

Electrolyte
(molten PbBr₂)

Figure 6.1   Electrolysis circuit and close-up of electrodes showing ions

by the cathode receive electrons from the cathode and form electrically neutral atoms. At the anode (positive electrode) negatively charged ions give up one or more of their electrons and also become electrically neutral atoms. The electrons leave the anode and travel to the power source. In electrolysis the electrolyte conducts electricity but decomposes as it does so.

In the electrolysis of lead bromide, for example, the positively charged lead ions receive electrons at the cathode and become lead atoms. They collect on the cathode then fall to the bottom of the container. At the same time, the negatively charged bromide ions release their electrons at the anode and become atoms which collect together to form brown bubbles of bromine gas that escape from the liquid.

This reaction can be written as the word equation

lead bromide → lead + bromine

The symbol equation for the reaction is

$PbBr_2$ (l) → Pb (1) + Br_2 (g)

An equation can be written for the reaction taking place at each electrode. Each equation is called a **half equation**.

The half equation for the reaction at the cathode is

$$Pb_2+ (l) 2e- \rightarrow Pb(l)$$

The half equation for the reaction at the anode is

$$2Br- (l) -2e \rightarrow Br_2 (g)$$

As in all equations the half equations must balance.

QUESTION

14 Balance these half equations
  (a) $Cl^- - e \rightarrow Cl_2$
  (b) $Cu^{2+} + e \rightarrow Cu$
  (c) $I^- - e \rightarrow I_2$

6.12 Summary

- Reactions can be represented by word equations (see p. 74).
- The formula of an ionic compounds can be worked out by studying the charges on the ions in the compound (see p. 75).
- The formula of covalent compounds can be worked out by studying how the elements share their electrons (see p. 77).
- Reactions can be represented by symbol equations (see p. 75).
- State symbols are also used in symbol equations (see p. 78).
- Symbol equations must be balanced (see p. 79).
- Ionic equations can be produced by the removal of spectator ions (see p. 80).
- Half equations are used to study the reactions taking place at the electrodes during electrolysis (see p. 81).

■ ⊻ **7** The periodic table

Objectives

When you have completed this chapter you should be able to:
- Identify **groups** and **periods** in the periodic table
- Understand how the atomic structure of the elements relates to their **positions** in the periodic table
- Describe some of the properties and uses of the **noble gases**
- Describe some of the properties and uses of the **alkali metals**
- Describe some of the properties and uses of the **alkaline earth metals**
- Describe some of the properties and uses of the **halogens**
- Describe some of the properties and uses of the **transition metals**.

7.1 Arranging the elements in order

The early alchemists knew nine elements They did not arrange them in order but associated them with objects in the Solar system. Gold was linked with the Sun, silver was linked with the Moon, iron was linked with Mars and lead with Saturn. By the early part of the nineteenth century over 50 elements had been discovered. They varied greatly in their properties and attempts were begun to put them into an order so that they could be easier to study.

Johan Dobereiner noticed that some groups of elements showed a gradation in a wide range of properties including reactivity. He noticed that the properties of calcium, strontium and barium showed a gradation and that the properties of two other groups of three elements showed a similar gradation. These groups were:

(1) sulphur, selenium and tellurium
(2) chlorine, bromine and iodine.

From his observations, he put forward his **Law of Triads** as a means of grouping the elements; however, other elements did not show this pattern and the Law of Triads did not become established.

In the early studies on the elements a unit called the atomic weight was used. This unit is no longer used today but it helped chemists in the past sort out the elements into order. In 1864 John Newlands arranged the elements in order using their atomic weights. He found that when he set out the ordered elements in columns of seven, some elements with similar properties occured in the same horizontal rows (Table 7.1).

For example fluorine and chlorine, lithium, sodium and potassium, magnesium and calcium occurred together. However other elements did not show similarities and many chemists considered the similarities that the table did show were just coincidences. Newlands called his arrangement **The Law of Octaves** after the seven notes of the musical scale. It was not seen as an important arrangement of the elements until after the work of Dimitri Mendeleev was published in 1869.

Mendeleev arranged the elements in order of their atomic weights but he also considered their **valency** which is the power with which they combine with other atoms. At the time, 63 elements were known and Mendeleev suspected that there were more to be discovered and left gaps for them in his table (Table 7.2).

Table 7.2 shows the periodic table that we use today. It is based on the table constructed by Mendeleev but includes a large number of elements that were not known in his time.

Although the table may look difficult to understand when first seen, its structure will become clearer if you think about the following points

- The elements are arranged **in order of their atomic numbers**, starting with hydrogen on its own at the top then moving across to helium at the top right. From helium you move to lithium on the left in the next row then move right again along the row. You can then follow the increase in atomic number by reading left to right along each line.
- The elements can be divided into two groups. They are the **metals** and **non-metals**. In the periodic table a line can be drawn step-wise from boron to astatine which separates the metals (on the left of the line) from the non-metals (on the right of the line).
- Some elements close to the line between metal and non-metals can also be grouped as **metalloids** or **semi-metals**. These elements are boron, silicon, germanium, arsenic, tellenium, antimony and selenium. They have some properties of metals and some properties of non-metals.
- The lines of elements you scanned as you looked at the atomic numbers of the elements are called **periods**. The first period contains hydrogen and helium. The second period contains the elements lithium to neon. There are seven periods. The last period begins with francium.

H	F	Cl
Li	Na	K
Ga	Mg	Ca
B	Al	Cr
C	Si	Ti
N	P	Mn
O	S	Fe

Table 7.1   The first three columns of elements, as set out by J.A.R. Newlands

Group	I Alkali metals	II Alkaline earth metals												III	IV	V	VI	VII Halogens	0 Noble gases
Period 1	1 H Hydrogen 1.0																		2 He Helium 4.0
2	3 Li Lithium 6.9	4 Be Beryllium 9.0												5 B Boron 10.8	6 C Carbon 12.0	7 N Nitrogen 14.0	8 O Oxygen 16.0	9 F Fluorine 19.0	10 Ne Neon 20.2
3	11 Na Sodium 23.0	12 Mg Magnesium 24.3												13 Al Aluminium 27.0	14 Si Silicon 28.1	15 P Phosphorus 31.0	16 S Sulphur 32.1	17 Cl Chlorine 35.5	18 Ar Argon 39.9
4	19 K Potassium 39.1	20 Ca Calcium 40.1	21 Sc Scandium 45.0	22 Ti Titanium 47.9	23 V Vanadium 50.9	24 Cr Chromium 52.0	25 Mn Manganese 54.9	26 Fe Iron 55.9	27 Co Cobalt 58.9	28 Ni Nickel 58.7	29 Cu Copper 63.5	30 Zn Zinc 65.4		31 Ga Gallium 69.7	32 Ge Germanium 72.6	33 As Arsenic 74.9	34 Se Selenium 79.0	35 Br Bromine 79.9	36 Kr Krypton 83.8
5	37 Rb Rubidium 85.5	38 Sr Strontium 87.6	39 Y Yttrium 88.9	40 Zr Zirconium 91.2	41 Nb Niobium 92.9	42 Mo Molybdenum 95.9	43 Tc Technetium (99)	44 Ru Ruthenium 101.1	45 Rh Rhodium 102.9	46 Pd Palladium 106.4	47 Ag Silver 107.9	48 Cd Cadmium 112.4		49 In Indium 114.8	50 Sn Tin 118.7	51 Sb Antimony 121.8	52 Te Tellurium 127.6	53 I Iodine 126.9	54 Xe Xenon 131.3
6	55 Cs Caesium 132.9	56 Ba Barium 137.3	57 ▲ La Lanthanum 138.9	72 Hf Hafnium 178.5	73 Ta Tantalum 181.0	74 W Tungsten 183.9	75 Re Rhenium 186.2	76 Os Osmium 190.2	77 Ir Indium 192.2	78 Pt Platinum 195.1	79 Au Gold 197.0	80 Hg Mercury 200.6		81 Tl Thallium 204.4	82 Pb Lead 207.2	83 Bi Bismuth 209.0	84 Po Polonium (210)	85 At Astatine (210)	86 Rn Radon (222)
7	87 Fr Francium (223)	88 Ra Radium (226)	87 ▲▲ Ac Actinium (227)	104 Unq Unnilquadium (261)	105 Unp Unnilpentium (262)	106 Unh Unnilhexium (263)													

▲ Lanthanide elements

58 Ce Cerium 140.2	59 Pr Praseodymium 140.9	60 Nd Neodymium 144.2	61 Pm Promethium (147)	62 Sm Samarium 150.4	63 Eu Europium 152.0	64 Gd Gadolinium 157.3	65 Tb Terbium 158.9	66 Dy Dysprosium 162.5	67 Ho Holmium 164.9	68 Er Erbium 167.3	69 Tm Thulium 168.9	70 Yb Ytterbium 173.0	71 Lu Lutetium 175.0

▲▲ Actinide elements

90 Th Thorium 232.0	91 Pa Protactinium (231)	92 U Uranium 238.1	93 Np Neptunium (237)	94 Pu Plutonium (242)	95 Am Americium (243)	96 Cm Curium (247)	97 Bk Berkelium (245)	98 Cf Californium (251)	99 Es Einsteinium (254)	100 Fm Fermium (253)	101 Md Mendelevium (256)	102 No Nobelium (254)	103 Lr Lawrencium (257)

Table 7.2 The periodic table with line showing division into metals and non-metals

- There are eight columns of elements which are numbered 1–0. Each column is called a **group**.
- The metals between group 2 and group 3 are called the **transition metals**. There are two sub-groups of transition metals which are placed below the main body of the table. They are:
  (1) the **lanthanides** or rare earth metals, which fit between barium and hafnium
  (2) the **actinides**, which fit between radium and unilquadium.

QUESTIONS

1 What are the elements in the third period of the periodic table?
2 What are the elements in the fifth group of the periodic table?
3 How does the metallic or non-metallic nature of the elements change as you move from left to right along a period?

7.2 Atomic structure and the periodic table

(i) The atoms in a period

If the atomic structure of the atoms in a period such as period 3 are examined they are found to show a trend as you move from left to right along the period. The number of electrons in the outer shell increases by one as you move from one element to the next (see Figure 7.1).

The table is known as the periodic table because as you examine the elements in the table you find elements with similar properties periodically. These elements are arranged in groups and form the larger columns in the table. For example, aluminium has three electrons in its outer shell and silicon has four.

The number of electrons in the outer shell of an atom in a group is the same

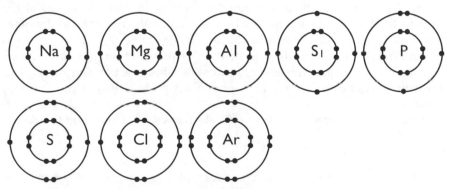

Figure 7.1 Atoms of sodium to argon just showing nucleus and electrons in outer shell

as the number of the group. For example, aluminium has three electrons in its outer shell and is in group 3. The exceptions are the elements in group 0. Each one has a full outer shell of electrons.

(ii) The atoms in a group

If the atomic structure of the atoms in a group such as group 1 is examined the elements will be seen to show these trends as you move from the top to the bottom of the group.

The size of the atoms of the elements increases and the densities of the elements also increases. The number of electrons in the outer shell remains the same for all the elements but the distance between these electrons and the nucleus increases as you move from one element to the one below it. This difference in distance affects metallic and non-metallic elements in different ways.

(iii) Metallic elements

When a metal atom takes part in a chemical reaction it loses its outer electrons and becomes a **metal ion**. A metal at the top of a group has a shorter distance between its nucleus and the electrons in its outer shell than a metal lower down the table. The force between the nucleus and the electrons in the outer shell of a metal at the top of a group is stronger than the force between the nucleus and the electrons in the outer shell of an element lower down the group. This difference in the strength of the force results in the metals lower down the group being more reactive than those higher up. because they can lose their electrons with greater ease. The reactivity of the metals increases as you move down the table.

(iv) Non-metallic elements

When the atom of a non-metallic element such as those in group 7 (the halogens) takes part in a chemical reaction it gains electrons in its outer shell and becomes a **non-metallic ion**. The non-metal atom at the top of the group has a shorter distance between its nucleus and outer electron shell than the atoms of the elements lower down the group. The force attracting the electrons to the outer shell is stronger in an element at the top of the table than in an element lower down. This results in the non-metallic elements at the top of the group being more reactive than the elements lower down because they can attract electrons with greater ease. The reactivity of non-metals increases as you move up the table.

For more about how atoms lose or gain electrons see Chapter 5 on Bonding.

QUESTION

4 How does the reactivity of the elements in group 1 compare with those in group 7? Explain your answer.

7.3 The melting and boiling points of some elements

The way the melting and boiling points of the metallic and non-metallic elements vary is described in detail for elements in group 1 (see p. 92), group 2 (see p. 97) and group 7 (see p. 99).

7.4 The position of hydrogen

A hydrogen atom has one electron in its outer electron shell like the atoms of the elements in group 1 (the alkali metals). It also forms ions with one positive electrical charge like the elements in the alkali metals. However, hydrogen is a gas at normal temperatures and pressures while the alkali metals are solids. Hydrogen also does not have any metallic properties (see p. 107) and usually forms covalent bonds (see p. 65) while metals form ionic bonds (see p. 60). As hydrogen has some properties of both the alkali metals and non-metals it is placed mid-way between these elements at the top of the periodic table. The preparation of hydrogen:

Hydrogen may be prepared in the laboratory as shown on p. 110. It is prepared industrially from brine as shown on p. 94.

Hydrogen is a colourless, non-poisonous gas which does not have a smell. It burns in air and can be detected by testing an ignition tube full with a flame. The hydrogen produces an explosive reaction which makes a squeaky pop.

QUESTION

5 Why is hydrogen not placed (a) in group 1, (b) with the non-metals?

7.5 Atoms and stability

The most stable or unreactive of elements are those which have their outer electron shells full of electrons. These elements form group 0 in the periodic table. They were discovered between 1894 and 1900 and are called the **noble** or **inert gases**. They were not found to take part in any chemical reactions until 1962. Today they are known to take part in a small number of reactions.

When elements take part in a chemical reaction they form full outer electron shells like the noble gases and their atoms become more stable. For example, the alkali metals lose an electron from their outer shell to expose a full electron shell beneath the one from which the electron has escaped.

When an atom of a halogen (see p. 99) takes part in a chemical reaction it gains an electron to form a full outer shell and gain the stability of an atom of the noble gases.

7.6 Group 0: the noble gases

Some physical properties of the noble gases are shown in Table 7.3.

(i) Reactions of the inert gases

None of the gases were found to react until 1962 when it was discovered that xenon reacted with fluorine to form compounds. Since then krypton has been found to form compounds with fluorine too and argon has been found to form compounds with boron fluoride.

(ii) The reason for the unreactivity

The noble gases have full outer electron shells which are extremely stable. Helium has two electrons in its outer shell and the other gases have eight electrons. Atoms of the noble gases do not join together like the atoms of other air gases such as oxygen or nitrogen (see pp. 127 and 123) but remain separate.

(iii) The properties and uses of the noble gases

Despite their chemical unreactivity the noble gases have a wide range of uses. In many cases it is their unreactivity that makes them particularly useful.

(a) Helium

This is the noble gas with the lowest melting and boiling points. It is used in cryogenics – low-temperature physics experiments. Helium, like hydrogen is less dense than air but unlike hydrogen is inflammable. It is used to safely raise weather balloons that carry instruments to measure atmospheric conditions and to provide the buoyancy in air ships. It is used in the breathing mixture for deep sea divers and as a pressurising gas in space rockets.

(b) Neon

Neon glows orange–red when a current of electricity is passed through it. Tubes of neon gas are lit by electricity in advertising signs to attract attention.

Element	Symbol	Density (g/cm^3)	Boiling point (°C)
helium	He	0.00008989	−269
neon	Ne	0.000899	−246
argon	Ar	0.001784	−185
krypton	Kr	0.003749	−152
radon	Rn	0.01009	−65

Table 7.3   Physical properties of the noble gases

(d) Argon

Argon is the most common noble gas in the air. It forms 1% of the atmosphere. Argon is used in light bulbs. It provides an inert atmosphere inside the glass which stops the metal filament burning out quickly.

(e) Krypton

Krypton is used in high-intensity lamps such as airport landing lights and in the lamp of a lighthouse. It is also used in fluorescent lamps and in gas-filled electronic devices.

(f) Xenon

Xenon is also used in high-intensity lamps and in photographic flash guns.

(g) Radon

Radon is the densest gas known. It is a radioactive element and does not have any uses. It is released from certain kinds of granite and if these are used for constructing buildings the gas may collect in quantities which are harmful to health. Today these buildings are checked for their radon content.

(iv) The noble gases and lasers

Noble gases are used to make laser beams. When a flash of light is introduced into a tube of helium and neon it raises the level of energy in the atoms of the gas and they release light too. The light from each atom makes other atoms release light until they form a powerful beam which is released from the tube. Weak laser light is used to read bar codes on items for sale in shops. Stronger lasers are used in eye surgery and in industry to drill holes in metal or weld pieces of metal together.

QUESTIONS

6 Which is (a) the least dense, (b) the most dense noble gas?
7 How does the density of the gas and the boiling points of the gases compare as you look down the group?
8 Why are the noble gases so unreactive?
9 Why is helium used instead of hydrogen in weather balloons and airships?
10 Why is argon used instead of air in light bulbs?

7.7 Describing chemical reactions

The chemical reactions of the elements in groups 1, 2 and 7 are described in the following part of the chapter. The reactions are described using word equations and symbol equations with state symbols and half equations as explained in Chapter 6, Representing reactions.

(i) Group I – the alkali metals

These metals are called alkali metals because they form **alkaline solutions** (see p. 153) with water.

The six elements in this group are lithium, sodium, potassium, rubidium, caesium and francium. The size of the atom changes down the group and this affects the reactivity of each element. Francium is also a radioactive element.

Metal	Symbol	Hardness (mohs)	Density g/cm^3	Melting point °C	Boiling point °C
lithium	Li	0.6	0.53	180	1336
sodium	Na	0.4	0.97	98	883
potassium	K	0.5	0.86	64	759
rubidium	Rb	0.3	1.53	39	700

Table 7.4 The properties of the first four alkali metals

The first three alkali metals have such low densities that they float on water. They are so soft that they can be cut with a knife. A freshly cut surface is shiny but it soon becomes dull and tarnished as the metal reacts with the air. Alkali metals are stored in oil to prevent their surfaces coming into contact with the air. The melting points and boiling points of alkali metals are low compared to other metals. For example, the melting points and boiling points of aluminium are 660°C and 2525°C; of calcium 840°C and 1493°C; and of iron 1535°C and 2865°C.

(a) The reaction of an alkali metal with water

Lithium fizzes with water while potassium melts in the heat of the reaction and the hydrogen produced bursts into flame. Rubidium and water explode!

(b) The reaction with water

lithium + water → lithium hydroxide + hydrogen
$2Li \ (s) + H_2O \ (l) \rightarrow 2LiOH \ (aq) + H_2 \ (g)$
sodium + water → sodium hydroxide + hydrogen
$2Na \ (s) + H_2O \ (l) \rightarrow 2NaOH \ (aq) + H_2 \ (g)$
Potassium + water → potassium hydroxide + hydrogen
$2K \ (s) + H_2O \ (l) \rightarrow 2KOH \ (aq) + H_2 \ (g)$

The hydroxide solutions formed in these reactions are strong alkaline solutions of pH 13–14.

(c) The reaction with oxygen

The metals burn brightly in oxygen. Sodium has a yellow flame and potassium has a lilac flame.

lithium + oxygen → lithium oxide

$4Li\ (s) + O_2\ (g) → 2Li_2O\ (s)$

sodium + oxygen → sodium oxide

$4Na\ (s) + O_2\ (g) → 2Na_2O(s)$

potassium + oxygen → potassium oxide

$4K\ (s) + O_2\ (g) → 2K_2O\ (s)$

The oxides are white powders. They are ionic solids. Each metal ion has one positive charge and each oxygen ion has two negative charges.

(d) The reaction with chlorine

The metals burn in chlorine. Lithium burns with a steady flame while potassium burns with the brightest flame of the three metals.

lithium + chlorine → lithium chloride

$2Li\ (s) + Cl_2\ (g) → 2LiCl\ (s)$

sodium + chlorine → sodium chloride

$2Na\ (s) + Cl_2\ (g) → 2NaCl\ (s)$

potassium + chlorine → potassium chloride

$2K\ (s) + Cl_2(g) → 2KCl\ (s)$

The metal chlorides produced in this reaction appear as a white smoke.

(e) The difference in reactivity and atomic structure

The activity between the alkali metal and water, oxygen and chlorine increases down the group. This is due to the increase in the size of the atom with more and more shells shielding the single electron in the outer shell from the force of attraction between it and the nucleus. With less and less force holding the electron in place the metals lower down the group more readily shed their electron in a chemical reaction to make a full outer electron shell which gives the ion the stability of a noble gas.

QUESTIONS

11 Would you expect caesium to be harder or softer than rubidium?

12 How would you expect the density to change in the two lowest members of the group?

13 What would you expect the melting points of caesium and francium to be? (*Note:* mercury is the only liquid metal at 20°C).

14 Would you expect caesium to react (a) more vigorously, or (b) less vigorously than rubidium?

15 Has a caesium atom (a) one more electron shell, or (b) one less electron shell than rubidium?

16 Is the force of attraction between the nucleus and the outer electron (a) stronger, or (b) weaker than in the rubidium atom?

17 Does the outer electron of a caesium atom escape (a) less easily, or (b) more easily than the outer electron in a rubidium atom?

18 Why does francium react more vigorously than caesium?

(f) The compounds of alkali metals

Sodium chloride

Sodium chloride occurs naturally in solid form as rock salt. This is also known as the mineral called halite. This mineral formed in vast quantities where an ancient sea evaporated. This occurred in an area that is now known as Cheshire in the United Kingdom. The huge deposit of rock salt was covered with other rock but today it is mined to provide sodium chloride for a wide range of uses.

Sodium chloride is also the most common salt in sea water. In hot coastal regions sodium chloride is extracted from sea water by running the sea water into shallow pans where the water evaporates leaving the salt behind (see Figure 10.7).

Uses of sodium chloride

(1) Salt is required by the body for the working of the nervous system, the movement of muscles, the production of hydrochloric acid in the stomach for digestion and for maintaining the health of all body cells.
(2) In the food industry, salt is used as a preservative and as a flavouring.
(3) Rock salt is used on the roads in winter to melt ice.
(4) Salt is made into a concentrated aqueous solution called brine. This:
  a is decomposed by electrolysis to make sodium hydroxide, chlorine and hydrogen which are used in industry
  b has carbon dioxide and ammonia added to it in the Solvay process to make sodium carbonate and sodium bicarbonate for use in industry.

The electrolysis of brine

Brine (a concentrated aqueous solution of sodium chloride) is decomposed by electrolysis in a **membrane cell** (Figure 7.2). This type of cell is replacing an

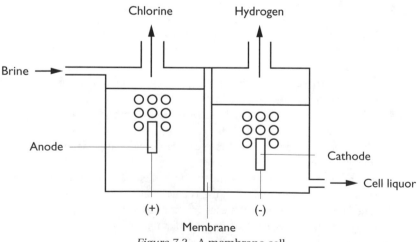

Figure 7.2 A membrane cell

older type of cell called a diaphragm cell because it is more efficient. Sodium hydroxide, chlorine and hydrogen are formed.

The equations for this reaction are:

sodium chloride + water → sodium hydroxide + chlorine + hydrogen

$2NaCl\ (aq) + 2H_2O\ (l) \rightarrow 2NaOH\ (aq) + Cl_2\ (g) + H_2\ (g)$

The cell is divided into two chambers by an asbetos partition called a membrane. This keeps the two gases separate as they are produced. Brine enters the chamber containing the anode. At this electrode chloride ions give up their electrons and form chlorine atoms. They join together in pairs to form molecules of chlorine gas:

$Cl^- \rightarrow Cl\ (g) + e^-$

$2Cl\ (g) \rightarrow Cl_2\ (g)$

The solution passes through the membrane into the chamber containing the cathode. Here hydrogen ions receive electrons from the cathode and form atoms which join together in pairs to form molecules of hydrogen gas:

$H^+ + e^- \rightarrow H\ (g)$

$2H\ (g) \rightarrow H_2\ (g)$

The two gases produced by electrolysis are collected separately from the top of the cell.

The hydroxide ions (OH^-) produced by the splitting of water by electolysis remain in the solution of sodium ions (Na^+) and chloride ions(Cl^-). This liquid is drawn out of the cell, and sodium chloride is separated from it by crystallisation and filtration to leave a solution of sodium hydroxide.

As both chlorine and alkali are produced by the electrolysis of brine the industry which makes these substances is called the chlor-alkali industry.

Uses of sodium hydroxide

Sodium hydroxide is used in a very wide range of industries. For example, it is used in chemical production, the making of artificial textile fibres, paper-making, processing metals and rubber and making dyes and bleaches. It is also an essential chemical in the production of soap.

Sodium hydroxide and soap

Soap is made from the fats and oils of animals and plants. These are first heated with water under pressure to raise the temperature and break them down into fatty acids and glycerol. Sodium hydroxide is added to react with the fatty acids in this boiling mixture and soap is produced.

A soap molecule has a head and a tail. The sodium ion is in the head. Carbon and hydrogen atoms from the fatty acids form the tail. The head is attracted to water and can be described as 'water-liking'. The tail is repelled by water and attracted to grease. It can be described as 'water-hating'.

Soap molecules bring water into greater contact with surfaces by reducing the surface tension, and clean surfaces by removing the grease.

The forces between the water molecules at the surface of a drop pull them together. This makes the surface behave like a skin and it holds the drop in a spherical shape. There is only a small area in contact between a water drop and a dirty surface. When soap is added to the water some soap molecules collect at the surface with their 'water-hating' tails sticking out. The 'water-liking' heads separate the water molecules in the surface. This reduces the forces of attraction between the water molecules and reduces the surface tension. The drop collapses onto the surface and increases its area of contact.

The tails of the soap molecules stick into the grease on the surface and lift it away. The heads of the soap molecules prevent the tiny balls of grease from joining together again and they flow easily away when the dirty water is poured away (Figure 7.3).

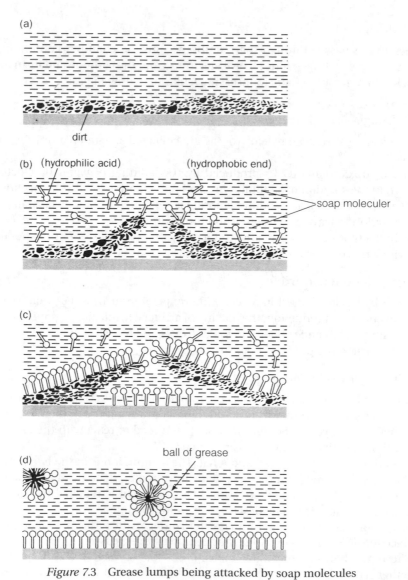

Figure 7.3 Grease lumps being attacked by soap molecules

Uses of sodium carbonate

Sodium carbonate is used for making glass, chemicals and detergents.

Uses of sodium hydrogen carbonate

Sodium hydrogen carbonate breaks down to carbon dioxide and steam when it is heated. It is called **baking soda**, and added to flour to make self-raising flour. When the flour is used in baking the carbon dioxide and steam lift the mixture in bread and cakes and give them a spongy texture.

The reaction is:

sodium hydrogen carbonate → sodium carbonate + carbon dioxide
+ steam

$$2NaHCO_3 \text{ (s)} \rightarrow Na_2CO_3 \text{ (s)} + CO_2 \text{ (g)} + H_2O \text{ (g)}$$

QUESTION

19 Assess the usefulness of sodium compounds.

(ii) Group 2 – the alkaline earth metals

These metals are called the alkaline earth metals because they form alkaline solutions when they react with water and they are found widely in many different kinds of rock on the Earth's surface.

The six elements in this group are beryllium, magnesium, calcium, strontium, barium and radium (Table 7.5). The size of the atom increases down the group as explained on p. 88, and this affects the reactivity of each element. Radium is also a radioactive element. Strontium has some isotopes which are radioactive.

Element	Symbol	Hardness (moh)	Density g/cm^3	Melting point °C	Boiling point °C
beryllium	Be	4.0	1.86	1280	2500
magnesium	Mg	2.0	1.75	650	1105
calcium	Ca	1.5	1.55	850	1440
strontium	Sr	1.8	2.60	770	1370

Table 7.5 The properties of the first four alkaline earth metals

The alkaline earth metals are not as reactive as the alkali metals, but they are always found combined with other elements. When alkaline earth metals are separated into their elemental form they are silvery-white solids but their surface soon forms an oxide layer by reacting with the oxygen in the air. Strontium and barium react so strongly with the air that they are stored in oil to keep the air from them.

(a) The reaction of an alkaline earth metal and water

Magnesium reacts very slowly with water and forms magnesium hydroxide and hydrogen:

magnesium + water → magnesium hydroxide + hydrogen
$Mg\ (s) + 2H_2O\ (l) \rightarrow Mg(OH)_2\ (aq) + H_2\ (g)$

If magnesium is heated in steam it glows brightly and forms white solid magnesium oxide and hydrogen:

magnesium + steam → magnesium oxide + hydrogen
$Mg\ (s) + H_2O\ (g) \rightarrow MgO\ (s) + H_2\ (g)$

Calcium reacts with water to form a cloudy solution of calcium hydroxide from which bubbles of hydrogen escape:

calcium + water → calcium hydroxide + hydrogen
$Ca\ (s) + 2H_2O\ (l) \rightarrow Ca(OH)_2\ (aq) + H_2\ (g)$

(b) The reaction of the alkaline earth metals and oxygen

Magnesium bursts into flame when heated in air and burns with a brilliant white light. It forms magnesium oxide:

magnesium + oxygen → magnesium oxide
$2Mg\ (s) + O_2\ (g) \rightarrow 2MgO\ (s)$

Calcium forms calcium oxide when heated in air:

calcium + oxygen → calcium oxide
$2Ca\ (s) + O_2\ (g) \rightarrow 2CaO\ (s)$

The vigorous reaction between alkaline earth metals and oxygen can be seen in a firework display (Figure 7.4). Magnesium provides the blinding white light, strontium produces crimson (red) colours and barium produces green colours.

Figure 7.4 Fireworks

20   Would you expect barium to be harder or softer than strontium?
21   How would you expect the density of barium to compare with the density of strontium?
22   Would you expect the melting point of barium to be (a) 970°C, (b) 803°C, (c) 729°C, (d) 453°C?
23   Would you expect strontium to react (a) less vigorously, or (b) more vigorously than calcium?
24   Arrange the atoms of these elements in order of size starting with the smallest – radium, calcium, barium, strontium.
25   Which of the six elements has the largest number of full electron shells in its atom?
26   Which of the six elements has an atom which has the strongest force between its outer electron shell and the nucleus?

(iii) Group 7 – the halogens

The word halogen comes from the Greek word for 'salt producer'. When halogens react with metals they form **salts**. The halogens are the most reactive non-metalic elements. The five elements in the group are fluorine, chlorine, bromine, iodine and astatine (Table 7.6). Fluorine is so reactive it will combine with nearly all other elements – often very violently. Astatine is a rare element and is radioactive.

Element	Symbol	Density g/cm^3	Melting point °C	Boiling point °C	Colour
fluorine	F	0.0016	−223	188	pale yellow
chlorine	Cl	0.0032	−103	−35	yellow-green
bromine	Br	3.119	−7	59	red-brown
iodine	I	4.94	114	184	purple-black

Table 7.6   The properties of the first four halogens

(a) The reactions of halogens and metals

Halogens react with metals to form **salts**.

For example sodium and chlorine react together to produce the white solid sodium chloride:

sodium + chlorine → sodium chloride
$2Na (s) + Cl_2 (g) \rightarrow 2NaCl (s)$

The salt is formed by ionic bonds The charge on the halogen ion is Cl⁻.

The halogens with higher atomic numbers react with less vigour than those with lower atomic numbers.

The equations for the reaction of chlorine and iron are:

iron + chlorine → iron(III) chloride
$2Fe (s) + 3Cl_2 (g) \rightarrow 2FeCl_3 (s)$

Aluminium can be heated in a current of chlorine to form white crystals of aluminium chloride:

Aluminium + chlorine → aluminium chloride

$2Al\ (s) + 3Cl_2\ (g) → 2AlCl_3\ (s)$

(b) Displacement

A halide is a compound of a halogen and one other element. For example, potassium chloride, KCl, is a compound of potassium and chlorine.

If a more reactive halogen is brought into contact with a halide made from a less reactive halogen, the less reactive halogen is displaced.

For example if chlorine gas is added to a solution of potassium bromide the following reaction takes place:

chlorine + potassium bromide → potassium chloride + bromine

$Cl_2\ (g) + 2KBr\ (aq) → 2KCl\ (aq) + Br_2\ (aq)$

Chlorine displaces bromine because it has a smaller nucleus with fewer shells shielding the power of attraction between the nucleus and the electrons in the outer shell. It pulls in the electron from the potassium ion more strongly than bromine. The bromine ions gain stability by forming covalent bonds and becoming diatomic bromine molecules:

chlorine + bromine ions → chlorine ions + bromine

$Cl_2\ (g) + 2Br^-\ (aq) → 2Cl^-\ (aq) + Br_2\ (aq)$

Chlorine also displaces iodine from potassium iodide:

chlorine + iodine ions → iodine + chlorine ions

$Cl_2\ (aq) + 2I^-\ (aq) → 2Cl^-\ (aq) + I_2\ (aq)$

Bromine has a smaller nucleus than iodine and a stronger power to attract electrons so it can displace iodine too:

bromine + iodine ions → bromine ions + iodine

$Br_2\ (aq) + 2I^-\ (aq) → 2Br^-\ (aq) + I_2\ (aq)$

(c) Halogens as oxidising agents

Substances which take up electrons in a reaction are called **oxidising agents**. As it takes up electrons the substance is reduced and the substance that loses electrons is oxidised (see p. 113).

The ability of halogens to take up electrons shows them to be oxidising agents. Fluorine is such a powerful oxidising agent that many of its reactions are explosive! Chlorine which is a less powerful oxidising agent than fluorine has many uses.

(d) Uses of the halogens

Chlorine

a Bleach

Chlorine removes the colour from litmus paper and other materials which contain a coloured dye. The molecule which gives the material its colour is large but only a part of it produces the colour. Chlorine reacts with this part and forms a molecule which is colourless.

Bleaches are used in paper-making and the textile industry.

Household bleach is used to remove stains and to kill bacteria. It is made by reacting chlorine with sodium hydroxide:

chlorine + sodium hydroxide → sodium chlorate (I) + sodium chloride
+ water

Cl_2 (aq) + 2NaOH (aq) → NaOCl (aq) + NaCl (aq) + H_2O (l)

b Herbicides (weedkillers)

(1) *Non-selective weedkillers*

These kill all plants they come into contact with and are used for clearing vegetation.

When chlorine is heated with sodium hydroxide a powerful oxidising agent called sodium chlorate (V) is formed:

chlorine + sodium hydroxide → sodium chlorate (V) + sodium chloride
+ water

$3Cl_2$ (g) + 6NaOH (aq) → $NaClO_3$ (aq) + 5NaCl (aq) + $3H_2O$ (l)

The weedkiller called paraquat is made by combining chlorine with an organic molecule. It is also harmful to humans if not handled carefully, and in large quantities it can enter food chains and cause widespread environmental damage.

(2) *Selective weedkillers*

These kill plants which have broad leaves such as nettles and dandelions but are harmless to narrow-leaved plants such as grass and cereals like wheat and oats. They can be sprayed on crops where they will kill the weeds but leave the crop plants unharmed. Selective weedkillers are made from chlorine and organic chemicals.

c Insecticides

Chlorine is combined with organic chemicals to make DDT. This powerful insecticide was used world-wide until it was found to enter food chains and cause far-reaching environmental damage. It is still used in the control of mosquitos that carry malaria but has been replaced in Europe and the USA by safer compounds.

d Plastics

Chlorine is reacted with ethene gas to make vinyl chloride and hydrogen chloride:

Cl_2 (g) + $CH_2 = CH_2$ (g) → $CHCl - CH_2$ (g) + HCl (g)

The vinyl chloride molecules are then linked together to form poly vinyl chloride (PVC, see Figure 7.5) which is the plastic used for windowframes, floor tiles and shoes.

e Hydrochloric acid

A major way of producing hydrochloric acid in the past was to burn hydrogen in chlorine. This produced hydrogen chloride which was dissolved in water:

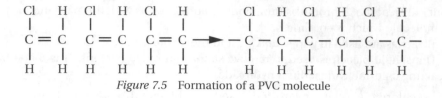

Figure 7.5 Formation of a PVC molecule

hydrogen + chlorine → hydrogen chloride
H_2 (g) + Cl_2 (g) → 2HCl (g)

hydrogen chloride $\xrightarrow[\text{water}]{\text{in}}$ hydrochloric acid

HCl (aq) → H^+ (aq) + Cl^- (aq)

Today most hydrogen chloride is made in the plastic industry. It is a by-product of a reaction in which molecules for making plastic are produced.

Iodine

Iodine is dissolved in alcohol to make tincture of iodine which is a common disinfectant. Iodine is also needed by the thyroid gland in the neck. It uses iodine to make a chemical which controls growth and activity. A lack of iodine in the food results in slow growth and mental activity in young people and a swelling in the neck, called a goitre, in older people. Iodine is added to table salt to prevent a lack of it in the diet. This kind of salt is called iodised salt.

Fluorine

The white coating of enamel on teeth can decay if there is a large amount of sugar in the diet. Fluorine ions reinforce the enamel against tooth decay by combining with the ions of calcium and phosphate. Fluorine is added to some water supplies and some toothpastes contain fluorine to increase the protection of tooth enamel. However too much fluorine makes the teeth go brown, hard and brittle so the addition of fluorine to water supplies and in toothpaste manufacture is carefully monitored to prevent damage to teeth.

(e) Hydrogen halides

Hydrogen forms covalent bonds with the halogens (Figure 7.6).

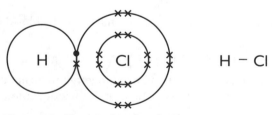

Figure 7.6 The covalent bond of hydrogen chloride

The solubility of the hydrogen halides in water is shown in Table 7.7. When the halides dissolve they form strong acids.

Formula	State	Colour	Solubility (%)	Ions
HCl	gas	none	42	H^+Cl^-
HBr	gas	none	68	H^+Br^-
HI	gas	none	70	H^+I^-

Table 7.7 The hydrogen halides

(f) Halides and photography

Silver halides are sensitive to light. Silver bromide and silver iodide crystals are used in films for black and white photography. When the light shines on them as the picture is taken some of the halides are destroyed and silver metal is released. It forms an image of the scene taken by the camera.

Silver bromide is used in the paper to make black and white prints. Some of the silver is released to form particles of metal as the image on the film is projected onto the paper.

QUESTIONS

27 From the information in Table 7.6 on p. 99, which
 (a) Halogens are gases at 20°C? Explain your answer.
 (b) Halogen is liquid at 20°C? Explain your answer.
 (c) Halogen is solid at 20°C? Explain your answer.
28 The atomic number of fluorine is 9, chlorine is 17 and bromine is 35. What does the atomic number of these three halogens tell you about their reactivity? Explain your answer.
29 If fluorine was brought into contact with a halide of chlorine. What would you expect to happen?

(iv) The transition metals

These elements are positioned between groups 2 and 3 in the periodic table.

Some of the metals have ions which produce coloured compounds. For example, the manganese ion $Mn2^+$ makes a pale pink colour, the iron Fe^{2+} ion produces a pale green colour, the iron Fe^{3+} ion produces a brown colour, the nickel 2+ ion produces a green colour and the copper Cu^{2+} ion produces a blue colour. Table 7.8 shows their melting points and densities.

	Sc	Ti	V	Cr	Mn	Fe	Co	Ni	Cu	Zn
Melting point	1541	1668	1890	1875	1244	1536	1493	1453	1088	419.6
Density g/cm³	9.98	4.51	6.11	7.19	7.44	7.87	8.90	8.908	9.94	7.1

Table 7.8 The melting points and densities of the transition metals

(a) The reactivity of the transition metals

The transition metals are much less reactive than the alkali metals and the alkaline earth metals. The reactivity of the metals is shown in the reactivity series on p. 108 (Table 8.1).

(b) Reaction with the oxygen in the air

Chromium and zinc react with oxygen in the air to form a very thin layer of oxide on their surfaces. This layer stops further reaction between the metals and oxygen. Iron does not react in dry air but if the air is damp a thin film of moisture forms on the iron's surface and rusting takes place.

Oxygen from the air dissolves in the water film. The oxygen takes part in a reaction with the water which draws electrons from the atoms of iron on the metal's surface. These atoms become ions and dissolve in the water. They join with the oxygen dissolved in the water to form brown flakes of iron oxide – rust:

$$\text{iron} + \text{oxygen} \rightarrow \text{iron (III) oxide}$$
$$4\text{Fe (s)} + 3\text{O}_2 \text{ (g)} \rightarrow 2\text{Fe}_2\text{O}_3 \text{ (s)}$$

Seawater and the slush of winter roads where ice has been melted by rock salt are especially corrosive to iron and steel. Structures and vehicles made of these metals can be dangerously weakened by rust.

(c) Rust prevention

Rust is prevented from forming in the following ways:

(1) **Paint** covers the surface and keeps oxygen and water from meeting the metal. It gives complete protection until the paint is scratched or chipped to the metal surface. Then rusting can begin.
(2) **Oil** can be used to cover the metal surfaces to keep oxygen and water away. Surfaces which are oiled to protect them, such as spades and garden tools, need re-oiling after use.
(3) When **zinc** is coated onto an iron surface crystals of an iron–zinc alloy form where the two metals meet. They hold the metals together. The outer surface of the zinc is in contact with the air and forms a layer of zinc oxide which prevents any corrosion. Iron treated with zinc in this way is called galvanised iron. If part of the zinc surface is chipped away oxygen will react with the zinc because zinc is more reactive than iron and the iron will not rust (see p. 116).
(4) **Alloys**: iron and steel can be prevented from rusting by mixing them with other metals to form alloys. A magnet does not rust because the iron is mixed with aluminium, cobalt or nickel. Stainless steel does not rust because the steel is mixed with nickel and chromium.

(d) Uses of some transition metals

Titanium

Titanium is very strong, corrosion resistant and light in weight. It is used for making spacecraft and artificial hip joints.

Iron

Iron can be treated to make cast iron which is very strong and wear resistant and is used for making car engines and manhole covers. Iron can also have a range of amounts of carbon added to it to make different kinds of steel. Hard steel has the most carbon in it and is used to make knives and scissors. Soft steels have the least amount of carbon in them and are used to make cans and car bodies.

Nickel

Nickel does not tarnish like many other metals. This means that its shiny metallic surface does not become dull by being in contact with the air. Nickel is mixed with other metals to make coins keep their shiny appearance.

Copper

Copper is a soft metal which can be pulled into a wire or shaped into tubes. It is a good conductor of heat and electricity and does not corrode. It is used for wiring the electrical circuits in all kinds of buildings and in the hot water systems of houses and flats.

Zinc

Zinc is used to galvanise iron and is mixed with copper to make the strong alloy called brass which is used in electrical plugs and sockets.

(e) Transition metals in jewellery

The major metals used in jewellery are transition metals from periods 5 and 6. They are silver and gold, which are placed beneath copper in the periodic table, and palladium and platinum, which are placed beneath nickel. Palladium is mixed with gold to make a less expensive alloy called white gold.

(f) Transition metals as catalysts

A catalyst is a substance which changes the speed of a reaction but is not changed by the reaction and can be used again. Catalysts are mostly used to speed up reactions They change the rate at which the reaction takes place. Catalysts are important in industry where they speed up the production of a wide range of products.

In the Haber process the catalyst is iron in the form of pellets. They are held in trays which are stacked in a 20 m-high tower. Hydrogen and nitrogen pass over them and join together to make ammonia.

Nickel is used to speed up the reaction between hydrogen and an oil to make margarine.

The catalytic converter on a car exhaust system speeds up the breakdown of harmful substances in the engine exhaust gases. It converts them to water and carbon dioxide. The catalysts in the converter are platinum and rhodium.

QUESTIONS

30 How do the melting points of the transition metals in period 4 change as you move along from left to right?

31 Which metal has a significantly different melting point from all the rest?
32 Which metal is the least dense? Another of its properties is strength. How are these two properties useful in the construction of fast flying aircraft?
33 Why should cars that have been driven through sea water be thoroughly washed?
34 Does chipped paintwork and a chipped galvanised coating cause the same problem? Explain.
35 How are the properties of copper exploited to provide a home with electricity and hot water?
36 How would the failure of a catalyst affect (a) margarine production, (b) car exhaust fumes?

7.8 Summary

- The periodic table is composed of elements arranged in periods and groups. (see p. 86).
- The elements are arrange in the periodic table in the order of their atomic numbers (see p. 85).
- The way an element takes part in chemical reactions is related to the atomic structure of that element (see p. 89).
- Hydrogen has some properties of both metals and non metals (see p. 89).
- The noble gases are the most stable of elements (see p. 89).
- The atoms of alkali metals lose an electron to gain the stability of the atoms of the noble gases (see p. 89).
- The alkali metals react with water, oxygen and chlorine (see p. 92).
- Sodium chloride has a wide range of uses (see p. 94).
- Sodium hydroxide is used in soap production (see p. 95).
- Alkaline earth metals react with water and oxygen (see pp. 97, 98).
- The atoms of halogens gain an electron to gain the stability of the atoms of the noble gases (see p. 89).
- Some halogens are capable of displacing other halogens from compounds (see p. 100).
- Halogens are oxidising agents (see p. 100).
- Chlorine is used for making bleach, pesticides and plastic (see pp. 100–1).
- The transition metals are found between groups 2 and 3 and have a wide range of uses (see p. 104).
- Transition metals react with oxygen (see p. 104).
- There are several ways to prevent the rusting of iron (see p. 104).

⬛▼ 8 Metals

Objectives

When you have completed this chapter you should be able to:
- Describe the physical properties of **metals**
- Understand the **reactivity series**
- Describe how some metals **react with oxygen**
- Describe how some metals **react with water**
- Describe how some metals **react with steam**
- Describe how some metals **react with dilute acids**
- Describe how some metal compounds are **decomposed by heat**
- Understand **displacement reactions**
- Understand **oxidation and reduction**
- Describe a **redox reaction**
- Explain how a knowledge of the reactivity series can prevent the **rusting of iron**.

The elements can be divided into two groups according to their general properties. These groups are metals and non-metals (see Chapter 9). The way the elements in periodic table are divided into these groups is described on p. 85.

8.1 The properties of metals

Metals:
- have high melting and boiling points (only mercury is a liquid at normal temperatures)
- are strong and can be shaped by hammering
- have a high density
- are good conductors of heat and electricity
- form positive ions
- form oxides and hydroxides which are bases which may dissolve in water to produce alkaline solutions
- form chlorides which are solids that are held together by ionic bonds.
- high in the reactivity series react with dilute acids.

8.2 The reactivity of metals

Metals vary in the way that they react with other substances. For example, potassium burns very vigorously in air while copper has to be heated strongly for it to react. When metals are made to react with air, water and dilute hydrochloric and sulphuric acids they all react in the same order. Potassium produces the most vigorous reactions, while other metals such as iron react less vigorously. The list of metals arranged in order of their reactivity starting with the most reactive, is called the **reactivity series**. Table 8.1 shows part of the reactivity series and how these metals react with oxygen, water and dilute acids.

(i) Examples of metals reacting with oxygen

Magnesium burns in oxygen with a brilliant white light which you must not look at directly. A white powder of magnesium oxide forms from this reaction:

magnesium + oxygen $\rightarrow$ magnesium oxide

$2Mg\ (s) + O_2\ (g) \rightarrow 2MgO\ (s)$

When aluminium is exposed to air a film of aluminium oxide forms on its surface (see also p. 183) without the spectacular burning process seen with magnesium:

Metal	Reaction with oxygen	Reaction with water	Reaction with acids
potassium sodium calcium magnesium aluminium zinc iron	burn in air form metal oxide	form metal hydroxide and hydrogen steam-oxide hydrogen	form metal salt and hydrogen
tin lead	does not burn – metal oxide	no reaction	no reaction
copper silver gold	no reaction		

Notes
- The metal at the top of each box in each column reacts most vigorously. The metal at the bottom of the box reacts least vigorously.
- The metals that react with hydrochloric acid form salts called **chlorides**. Metals that react with sulphuric acid form salts called **sulphates**.
- Nitric acid is not used to compare metals in the reactivity series.
- Gold only reacts with one substance. It is called **aqua regia**, and is a mixture of concentrated hydrochloric and nitric acids.
- The non-metal **carbon** fits in the table between aluminium and zinc.
- The non-metal **hydrogen** fits in the table between lead and copper.

Table 8.1 The reactivity series

aluminium + oxygen → aluminium oxide

$4Al\ (s) + 3O_2\ (g) → 2Al_2O_3\ (s)$

QUESTION

1 The formula for zinc oxide is ZnO. Produce a balanced symbol equation for the reaction between zinc and oxygen.

(ii) Examples of metals reacting with water

Metals at the top of the reactivity series react with cold water.

If a small piece of sodium is placed on the surface of water it rushes about and the heat produced by the reaction melts the metal. Eventually all the metal reacts with the water to produce sodium hydroxide and hydrogen:

sodium + water → sodium hydroxide + hydrogen

$2Na\ (s) + 2H_2O\ (l) → 2NaOH\ (aq) + H_2\ (g)$

Calcium dissolves in water to form calcium hydroxide and hydrogen:

calcium + water → calcium hydroxide + hydrogen

$Ca\ (s) + H_2O\ (l) → Ca\ (OH)_2\ (aq) + H_2\ (g)$

Magnesium, zinc and iron react slowly with water but react more quickly with steam.

(iii) Reaction of metals with steam

The metal is set up in the apparatus shown in Figure 8.1.

When the metals react with steam they form the metal oxide and hydrogen:

magnesium + steam → magnesium oxide + hydrogen

$Mg\ (s) + H_2O\ (g) → MgO\ (s) + H_2\ (g)$

zinc + steam → zinc oxide + hydrogen

$Zn\ (s) + H_2O\ (g) → ZnO\ (s) + H_2\ (g)$

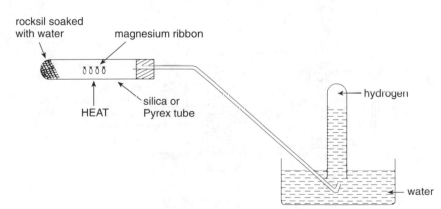

Figure 8.1 Heating a metal in steam

The reaction between iron and steam is a reversible reaction (see p. 75):

iron + steam ⇌ iron (II) diiron (III) oxide + hydrogen
$3Fe\ (s) + H_2O\ (g) ⇌ Fe_3O_4\ (s) + 4H_2\ (g)$

(see p. 75)

QUESTION

2 Potassium (K) reacts with water in a similar way to sodium. Construct a word equation and a balanced symbol equation for this reaction.

(iv) Reaction of metals with dilute acids

Sodium produces an explosive reaction with dilute acids:

sodium + hydrochloric acid → sodium chloride + hydrogen
$2Na\ (s) + 2HCl\ (aq) → 2NaCl\ (aq) + H_2\ (g)$

magnesium dissolves quickly in dilute hydrochloric acid.
 Zinc and sulphuric acid are used to produce hydrogen for collection in the laboratory using the apparatus shown in Figure 8.2:

zinc + sulphuric acid → zinc sulphate + hydrogen
$Zn\ (s) + H_2SO_4\ (aq) → ZnSO_4\ (aq) + H_2\ (g)$

Copper is below hydrogen in the reactivity series (see note to Table 8.1, p. 108) and cannot displace it from dilute acids like the metals above hydrogen can.

QUESTION

3 Write a word equation and balanced symbol equation for the reaction between each of these metals and sulphuric acid – (a) sodium, (b) calcium.

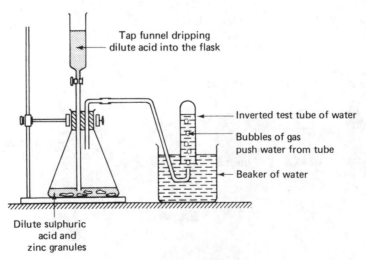

Tap funnel dripping
dilute acid into the flask

Inverted test tube of water

Bubbles of gas
push water from tube

Beaker of water

Dilute sulphuric
acid and
zinc granules

Figure 8.2 Preparation of hydrogen

(v) The effect of heat on metal compounds

Metal chlorides are not decomposed by heat but some metal oxides, hydroxides, carbonates and sulphates are decomposed by heat. The way that they decompose is related to their positions in the reactivity series. Where metal salts are decomposed by heat the salts of metals lower in the reactivity series decompose more readily than the salts of those higher up the series.

(a) Oxides

Gold does not form an oxide but silver oxide decomposes to form the metal and oxygen. The oxides of other metals shown in the series do not decompose.

(b) Hydroxides

Silver and gold do not form hydroxides but the hydroxides of the metals copper to calcium decompose to form the metal oxide and water. Sodium and potassium hydroxides do not decompose.

(c) Carbonates

Gold and silver carbonates decompose to form the metal, oxygen and carbon dioxide. The carbonates of copper to calcium decompose to form the metal oxide and carbon dioxide. Sodium and potassium carbonates do not decompose.

(d) Sulphates

The sulphates of gold and silver decompose to the metal, oxygen and sulphur trioxide. Some sulphates also decompose to sulphur dioxide. Sodium and potassium sulphates do not decompose.

QUESTIONS

4 Produce a word and balanced symbol equation for the effect of heat on (a) silver oxide (Ag_2O), (b) calcium hydroxide ($Ca(OH)_2$), (c) copper carbonate ($CuCO_3$).
5 Produce a table entitled 'The effect of heat on metal compounds', showing the information in this section.

(vi) Displacement reaction

In a displacement reaction a less reactive element in a compound is displaced (and replaced) by a more reactive one.

(a) The displacement of hydrogen

Hydrogen occupies a low position in the reactivity series. The more reactive metal elements above it can displace it from water, steam and dilute acids. The less reactive metal elements below it cannot displace it.

(b) The displacement of metals

Metal salts in aqueous solution

If a piece of metal which is more reactive than the metal forming the salt is added to the solution the metal displaces the less reactive metal from its salt.

EXAMPLE

When a piece of zinc is added to copper sulphate solution the following reaction takes place and copper is deposited on the piece of zinc:

zinc + copper sulphate → zinc sulphate + copper

$$Zn\ (s) + CuSO_4\ (aq) \rightarrow ZnSO_4\ (aq) + Cu\ (s)$$

The reaction is exothermic (see p. 5).

QUESTIONS

6 Which of these metals could displace iron from its salt in aqueous solution: lead, calcium, copper, zinc, magnesium, tin? Explain your answer.
7 If you have studied ionic equations (see p. 80) identify the spectator ions in the reaction between zinc and copper sulphate and produce an ionic equation for this reaction.
8 Is heat taken in or given out when displacement reactions take place? Explain your answer.

Displacement of a molten metal salt

If a solid salt of a less reactive metal is heated with a more reactive metal the less reactive metal is displaced.

EXAMPLE

iron + copper (II) bromide → iron (II) bromide + copper

$$Fe(s) + CuBr_2\ (s) \rightarrow FeBr_2\ (s) + Cu\ (s)$$

Displacement from metal oxides

When an oxide of a less reactive metal is heated with a more reactive metal the less reactive metal is displaced as the more reactive metal combines with oxygen. This type of reaction can be thought of as a **competition reaction** where the two metals compete to combine with oxygen. In competition reactions the 'winning' metal is the more reactive one and the 'losing' metal is the less reactive one.

The most spectacular competition reaction takes place between iron oxide and aluminium. The equations for this reaction are:

iron (III) oxide + aluminium → aluminium oxide + iron

$$Fe_2O_3\ (s) + Al\ (s) \rightarrow Al_2O_3\ (s) + 2Fe\ (s)$$

The two reactants are in powder form and the heat to start the reaction in a laboratory demonstration (see Figure 8.3) is provided by a burning piece of magnesium. The reaction produces large amounts of heat.

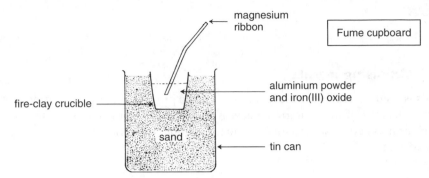

Figure 8.3 Thermit process laboratory set up

This reaction is also known as the **thermit process**, and is used to repair damaged railway lines

QUESTION

9 The reactivity series can be used to predict what will happen when certain metals and metallic compounds are brought together. Predict what will happen when (a) copper is added to silver nitrate solution, (b) silver is added to copper sulphate solution, (c) magnesium is added to copper sulphate solution, (d) zinc is added to magnesium sulphate solution. In each case, explain your answer.

8.3 Oxidation and reduction

The thermit process is a spectacular example of two processes called oxidation and reduction.

(i) Oxidation

Oxidation occurs when an atom of an element combines with oxygen. For example, in the thermit process the aluminium receives oxygen from the iron oxide. The oxidation process actually means more than just a combination with oxygen. It also means that an atom, ion or molecule may also lose electrons. For example in the thermit process, the aluminium atom loses three electrons to form an aluminium ion as the ionic equation shows:

$$Fe^{3+} + Al \rightarrow Fe + Al^{3+}$$

Oxidation also means the loss of hydrogen from an ion or molecule. For example, when methane burns in air the following reaction takes place:

methane + oxygen $\rightarrow$ carbon dioxide + water
$$CH_4 \text{ (g)} + O_2 \text{ (g)} \rightarrow CO_2 \text{ (g)} + H_2O \text{ (l)}$$

The hydrogen atoms are removed from the carbon atom in a methane molecule.

10 Summarise the different ways in which oxidation can be defined.

(ii) Oxidising agents

This is a substance that provides oxygen for the production of a compound, takes away electrons from an atom or takes away hydrogen from an ion or a molecule.

Examples of oxidising agents are the non-metals oxygen, chlorine and the acid nitric acid.

(iii) Reduction

In a reduction reaction the atoms or ions of an element in a compound lose oxygen. In the thermit process the iron (III) ions lose oxygen. This process is the opposite of oxidation. In a reduction reaction an atom may also gain electrons. In the thermit process the iron ions gain electrons and are reduced:

$$Fe^{3+} + 3e^- \rightarrow Fe$$

In a reduction reaction an element or compound can gain hydrogen. For example in the Haber process (p. 237) nitrogen combines with hydrogen to form ammonia:

nitrogen + hydrogen → ammonia
$$N_2 + 3H_2 \rightarrow 2\,NH_3$$

11 Summarise the different ways in which reduction takes place.

(iv) Reducing agents

This is a substance that takes away oxygen from a compound, gives electrons to an ion or gives hydrogen to an ion or molecule. Examples of reducing agents are metals and the non-metals carbon, hydrogen and the compound carbon monoxide.

(v) Redox reaction

Oxidation and reduction processes nearly always take place in the same chemical reactions. These reactions are called redox reactions. The thermit process and the displacement reactions (see p. 111) are examples of redox reactions.

12 The reaction between copper oxide and hydrogen is an example of a redox reaction:

copper oxide + hydrogen → copper + water
$$CuO\ (s) + H_2\ (g) \rightarrow Cu\ (s) + H_2O\ (l)$$

Which substance is oxidised and which is reduced? Explain your answer.

8.4 The position of hydrogen and carbon in the reactivity series

The positions of hydrogen and carbon in the reactivity series have been worked out by reacting each one with the oxides of the metals in the the reactivity series. When hydrogen is heated with the metal oxide, as shown in Figure 8.4, lead oxide and copper oxide are reduced, as the following equations show:

lead oxide + hydrogen → lead + water
PbO (s) + H_2 (g) → Pb (s) + H_2O (l)
copper (II) oxide + hydrogen → copper + water
CuO (s) + H_2 (g) → Cu (s) + H_2O (l)

When hydrogen is heated with iron oxide a reversible reaction takes place:

$$Fe_3O_4 + 4H_2 \rightleftharpoons 3Fe + 4 H_2O$$

Hydrogen cannot completely reduce the oxide so iron represents the point in the table at which reduction with hydrogen reaches its limit.

Metals above it in the reactivity series are not reduced by hydrogen.

Carbon in powder form can also reduce copper and lead oxides to form carbon dioxide and the metal. With very strong heat carbon can reduce zinc and iron oxides:

zinc oxide + carbon → zinc + carbon dioxide
$2 ZnO$ (s) + C (s) → $2 Zn$ (s) + CO_2 (g)
iron oxide + carbon → iron + carbon dioxide
$2Fe_2O_3$ (s) + $3C$ (s) → $4Fe$ (s) + $3CO_2$ (g)

QUESTIONS

13 Why is carbon placed higher in the reactivity series than hydrogen?
14 Produce balanced symbol equations for the reaction between (a) copper oxide (CuO) and carbon (C), (b) lead oxide (PbO) and carbon (C).

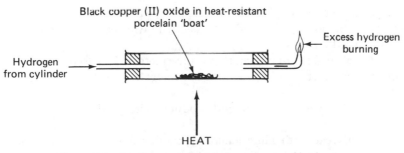

Figure 8.4   Heating metal oxide in a stream of hydrogen

8.5 Rust prevention and the reactivity series

Oxygen in the air reacts with many metals. It causes a freshly cut surface of sodium metal to become tarnished in minutes. In damp conditions it causes iron to rust. The ionic equation for the rusting of iron is:

$$Fe \rightarrow Fe^{3+} + 3e^-$$

(i) Sacrificial protection

In the rusting process atoms of irons lose three electrons and are oxidised. The rusting of a piece of iron can be prevented by attaching a lump of zinc or magnesium to it. This is done to iron used for underwater pipes and to the steel sheets (which contain iron) that are used to make the hulls of ships.

Zinc and magnesium are more reactive than iron and are oxidised more readily. The reaction shown for the ionic equation for rusting does not take place as the attacking chemicals react with the more reactive zinc and the iron is preserved. The magnesium or zinc blocks gradually dissolve in the water. They are said to be sacrificed for the iron and the process is called **sacrificial protection**. If the metal blocks are not replaced when they are used up the iron then begins to rust.

(ii) Galvanising

In this process, iron is coated (or plated) with zinc. As the zinc is more reactive than the iron it corrodes first and prevents the iron from rusting. It does this by supplying any attacking chemicals with electrons. If a galvanised surface is scratched the iron beneath does not rust because the zinc takes part in the reaction with attacking chemicals.

QUESTIONS

15 Why can magnesium and zinc be used to prevent rusting? Explain your answer.
16 Compare sacrificial protection with galvanising.

8.6 Summary

- Metals have a range of physical properties (see p. 107).
- Metals can be arranged in order of reactivity with oxygen, water and dilute acids (see p. 108).
- Some metals react with oxygen to produce metal oxides (see p. 108).
- Some metals react with water to produce metal hydroxides and hydrogen (see p. 109).
- Some metals react with steam and produce metal oxides and hydrogen (see p. 109).
- Some metals react with dilute acids and produce metal salts and hydrogen (see p. 110).

- Some metal compounds are decomposed by heat (see p. 111).
- In a displacement reaction a less reactive element in a compound is replaced by a more reactive one (see p. 111).
- Oxidation occurs when a substance receives oxygen, loses electrons or hydrogen (see p. 113).
- Reduction occurs when a substance loses oxygen, gains electrons or hydrogen (see p. 114).
- Oxidation and reduction take place at the same time in redox reactions (see p. 114).
- Zinc and magnesium can be used to prevent the rusting of iron (see p. 116).

■ ☑   **9** Non-metals

Objectives

When you have completed this chapter you should be able to:
- Describe the **physical properties** of non-metals
- Describe the **allotropes of carbon** and the uses of carbon
- Describe the reactions of **carbon and oxygen**
- Describe the test for **carbon dioxide**, its preparation, properties and uses
- Explain how carbon moves through the **carbon cycle**
- Understand why **nitrogen is an unreactive gas**
- Describe how nitrogen moves through the **nitrogen cycle**
- Describe the preparation and properties of **ammonia**
- Describe the preparation and properties of **oxygen**
- Understand how **fire is controlled**
- Describe the **extraction of sulphur**
- Identify the **allotropes of sulphur**
- Describe the reactions and uses of **sulphur and sulphur dioxide**
- Describe the preparation and properties of **chlorine**
- Describe the preparation and properties of **hydrogen chloride**
- Arrange some of the non-metals in an **order of reactivity**.

The elements in the periodic table can be divided into two groups – the metals (see Chapter 8) and the non-metals. You can find the non-metals in the periodic table in Table 7.2 p. 86. They are on the right of the step like thick line on the right hand side of the table.

9.1 The properties of non-metals

Non-metals:

- have low melting and boiling points (some are gases at normal temperatures)
- are brittle, the solids break up if they are hammered
- have a low density
- are poor conductors of heat and electricity (graphite, a form of carbon is an exception)
- form negative ions

- form oxides that are neutral and insoluble and acidic and soluble
- form chlorides which are liquids or gases
- do not react with dilute acids.

QUESTION

    1 How do non-metals compare with metals? (Look on p. 107 to help you.)

9.2 The non-metals

The non-metals in this chapter are considered as they are arranged in the peri-odic table – carbon, nitrogen, oxygen, sulphur, chlorine, bromine and iodine.

(i) Carbon

There are three forms or allotropes of pure carbon. They are diamond, graphite and fullerene which contains between 30 and 70 carbon atoms linked to form a sphere like shape. The fullerene with the most sphere-like shape has 60 carbon atoms and is called buckminster fullerene (see Figure 9.1).

    The sphere-like structures of the fullerenes are known as 'bucky balls'.

    Impure forms of carbon are coal, coke, charcoal and soot.

(a) Carbon and oxygen

When carbon is heated in oxygen or where the supply of air is not limited carbon dioxide is formed:

    carbon + oxygen → carbon dioxide
    $C (s) + O_2 (g) → CO_2 (g)$

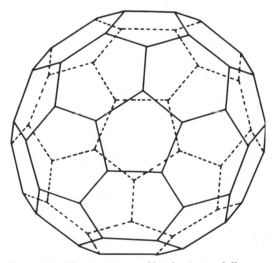

Figure 9.1 The structure of buckminster fullerene

Carbon dioxide makes up 0.033% of the atmosphere.

When carbon is heated with a limited supply of air carbon monoxide is produced:

$$C \text{ (s)} + O_2 \text{ (g)}$$

This gas affects the blood in the following way. The blood contains a red pigment, haemoglobin, which combines with oxygen to form oxyhaemoglobin. This compound transports oxygen around the body and delivers it to all parts that need it. Carbon monoxide combines with haemoglobin to form carboxyhaemoglobin. This compound is much more stable than oxyhaemoglobin and prevents oxygen being transported by haemoglobin molecules. As a result of this the body is starved of oxygen and dies.

QUESTION

2 Why is carbon monoxide a lethal gas?

(b) Carbon and water

Carbon does not react with cold water but it reacts with steam to form carbon monoxide and hydrogen:

carbon + steam → carbon monoxide + hydrogen
$$C \text{ (s)} + H_2O \text{ (g)} \rightarrow CO \text{ (g)} + H_2 \text{ (g)}$$

(c) Carbon and acids

Carbon does not react with dilute acids. Carbon reacts with hot concentrated nitric acid and sulphuric acid to form carbon dioxide.

(d) Carbon and metal oxides

Carbon reduces metal oxides below it in the reactivity series (see p. 108).

QUESTION

3 If you have studied the metals in Chapter 8, compare the reactivity of carbon with the metals that are (a) low, and (b) high in the reactivity series.

(e) Carbon dioxide

Laboratory preparation

Although carbon dioxide is slightly soluble it can also be collected over water like hydrogen. It can be prepared in the laboratory using the apparatus shown in Figure 9.2.

The calcium carbonate reacts with the hydrochloric acid to produce carbon dioxide, calcium chloride and water:

$$CaCO_3 \text{ (s)} + HCl \text{ (aq)} \rightarrow CO_2 \text{ (g)} + CaCl_2 \text{ (aq)} + H_2O \text{ (l)}$$

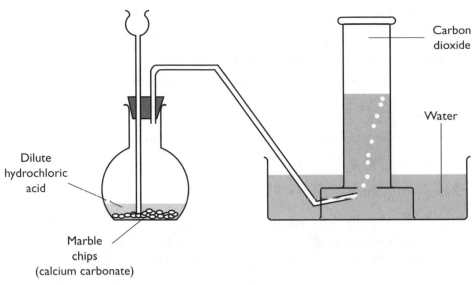

Figure 9.2 Preparation of carbon dioxide from calcium carbonate and hydrochloric acid

Properties

Carbon dioxide is a colourless, odourless gas. Only magnesium can burn in it. Carbon dioxide sublimes at −78°C and forms a solid known as dry ice. If this solid is warmed above −78°C it sublimes and forms gaseous carbon dioxide again.

Test for carbon dioxide

Carbon dioxide can be detected by bubbling the gas through lime water where it reacts with the calcium hydroxide in the solution to produce a cloudy precipitate of calcium carbonate:

$$CO_2 \text{ (g)} + Ca(OH)_2 \text{ (aq)} \rightarrow CaCO_3 \text{ (s)} + H_2O \text{ (l)}$$

If more carbon dioxide is passed into the lime water soluble calcium hydrogen carbonate forms and the precipitate disappears:

carbon dioxide + calcium carbonate + water
→ calcium hydrogen carbonate
$$CO_2 \text{ (g)} + CaCO_3 \text{ (s)} + H_2O \text{ (l)} \rightarrow Ca(HCO_3)_2$$

QUESTION

4 How good is the test for carbon dioxide if you left carbon dioxide bubbling through lime water before you looked at it? Explain your answer.

Carbon dioxide and water

Carbon dioxide takes part in a chemical reaction with water. It forms carbonic acid which has a pH of 5 or 6. The reaction is reversible:

carbon dioxide + water $\rightleftharpoons$ carbonic acid

CO_2 (g) + H_2O (l) $\rightleftharpoons$ H_2CO_3 (aq)

The oxidising properties of carbon dioxide

In a blast furnace (see p. 224) carbon dioxide oxidises carbon to form carbon monoxide:

C (s) + CO_2 (g) $\rightarrow$ $2CO$ (g)

When magnesium is burnt in carbon dioxide it is oxidised to magnesium oxide:

CO_2 (g) + $2Mg$ (s) $\rightarrow$ $2MgO$ (s) + C (s)

(f) Uses of carbon

Carbon in the form of:

- diamonds is used as jewellery and in cutting and drilling equipment
- graphite is used in pencil leads where it is mixed with clay; it is also used as a lubricant
- coke is used to extract iron form its ore
- charcoal is used in a wide range of filters from some of those used in gas masks to filters in aquarium tanks.

Carbon dioxide is used:

- to make the bubbles in fizzy drinks
- in fire extinguishers that are used to put out electrical fires; when the gas is released from the fire extinguisher it cascades over the fire and smothers it in a blanket of non-flammable material which displaces oxygen from the fire. As a consequence of this the fire goes out
- to create layers of 'smoke ' on stage in theatrical productions and pop music concerts. Dry ice is put in water to produce the white smoke that pours over the stage.

QUESTION

> 5 What physical property do you think carbon dioxide has that makes it particu-
> larly useful for extinguishing electrical fires?

(g) The carbon cycle

Life on earth is based on carbon. This means that the molecules which form living things have large numbers of carbon atoms linked together. Plants take in carbon in the form of carbon dioxide during the process of photosynthesis. The reaction which summarises photosynthesis is:

carbon dioxide + water $\rightarrow$ glucose + oxygen

$6CO_2$ (g) + $6H_2O$ (l) $\rightarrow$ $C_6H_{12}O_6$ (aq) + $6O_2$ (g)

Glucose is an organic compound. It may be used as a fuel in respiration or converted into other organic compounds to enable the plant to grow and continue to survive.

About 300 million years ago many plants lived in swamps that covered huge areas. When the plants died they fell beneath the waters of the swamps and did not decay but fossilised to form **coal**. The geological period in which this took place is now known as the Carboniferous Period. When the planktonic animals and plants in ancient seas died and sank to the sea bed they did not decay completely as they were covered by sedimentary rock but formed **oil** and **natural gas**.

Carbon in plants passes to herbivorous animals then carnivorous animals along the food chains. In each organism some of the carbon is released in the process of respiration as carbon dioxide. The remaining carbon is released when the organism dies and decays as microorganisms feed and respire.

Figure 9.3 shows how carbon moves through the environment.

QUESTIONS

6 Describe three paths taken by a carbon atom through the carbon cycle.
7 Which processes in the carbon cycle are most vital to maintaining life on Earth?
8 How do you think the carbon cycle looked before there were humans using industrial processes?

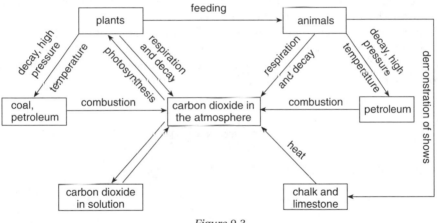

Figure 9.3

(ii) **Nitrogen**

Nitrogen is a colourless gas without a smell. It makes up 78% of the atmosphere. It is only slightly soluble in water and is a comparatively unreactive gas. This unreactivity is due to the three covalent bonds which bind a pair of nitrogen atoms together. A pair of nitrogen atoms make a nitrogen molecule. Molecules made of two atoms are called diatomic molecules. The nitrogen in the atmosphere is in the form of diatomic molecules.

QUESTION

9 Figure 9. 4 shows the electrons in the outer shell of a nitrogen atom. Eight are needed for the atom to be stable. Show how a pair of nitrogen atoms join

together and share their electrons. (Use the information on p. 68 about oxygen to help you answer.)

Figure 9.4   The outer shell of two nitrogen atoms

(a) Nitrogen and oxygen

Nitrogen and oxygen do not react at normal atmospheric temperatures but they react in the heat of a combustion engine of a car or truck to produce nitric oxide (NO) and nitrogen dioxide (NO_2) which are sometimes referred to as nitrogen oxides (NOx).

When a bolt of lightning shoots through the air it heats the air gases and causes nitrogen and oxygen to react to produce nitrogen oxides. These react with water vapour in the air to produce nitric acid. When the nitric acid reaches the ground it reacts with minerals which form the soil and nitrates are produced. Nitrates are taken up by plants and used to make proteins. Proteins are used to form the structure of plant and animal bodies. The way nitrogen passes between the living and non-living part of the environment is shown in Figure 9.5.

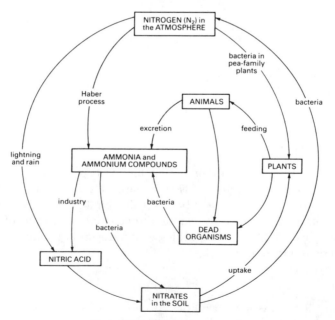

Figure 9.5   The nitrogen cycle

(b) Nitrogen and water

Nitrogen only dissolves slightly in water.

(c) Nitrogen and hydrogen

Nitrogen and hydrogen gas are heated under pressure in the Haber process to produce ammonia (see p. 238).

(see p. 238)

QUESTIONS

10 Compare the reactivity of nitrogen with the reactivity of carbon earlier in this chapter.

11 Why do you think there is a difference in reactivity?

(iii) Ammonia

Ammonia may be prepared in the laboratory using the reactants calcium hydoxide and ammonium chloride. The apparatus for preparing ammonia is shown in Figure 9.6.

The reaction which takes place is:

calcium hydroxide + ammonium chloride → calcium chloride + water
+ ammonia

$$Ca(OH)_2 \text{ (s)} + 2NH_4CL \text{ (s)} \rightarrow CaCl_2 \text{ (s)} + 2H_2O \text{ (g)} + NH_3 \text{ (g)}$$

It can be seen from the state symbols that both water and ammonia are in gaseous form. The water is removed as it passes though the calcium oxide in the drying tower.

(a) Properties of ammonia

Ammonia is a colourless gas but has a very strong smell. It is present in the urine of babies and is a major contributor to the smell of wet nappies. Ammonia is less dense than air.

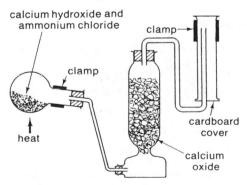

calcium hydroxide and
ammonium chloride

clamp

clamp

heat

cardboard
cover

calcium
oxide

Figure 9.6 Laboratory preparation of ammonia

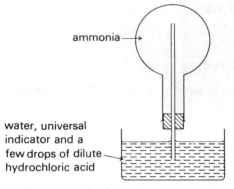

ammonia

water, universal
indicator and a
few drops of dilute
hydrochloric acid

Figure 9.7 The fountain experiment

(b) Ammonia and water

Ammonia is very soluble in water and is used in the fountain experiment shown in Figure 9.7.

When water enters the flask containing dry ammonia the gas dissolves and this reduces the pressure of the gas inside the flask. The difference in gas pressure between the ammonia in the flask and the air pressure pushing on the water in the trough causes the water to flow up into the flask. As ammonia is so soluble the pressure rapidly drops inside the flask and the water gushes in and makes a fountain.

When ammonia dissolves in water it makes an alkaline solution. The pH of the solution is about 10.

Ammonia reacts with water to form ammonium and hydroxide ions:

$$NH_3 \text{ (g)} + H_2O \text{ (l)} \rightarrow NH^{4+} \text{ (aq)} + OH^- \text{ (aq)}$$

Ammonia solution is used to identify metal ions in aqueous solutions. For example, a solution containing iron (II) ions produces a green precipitate while iron (III) ions produce a brown precipitate and copper produces a blue precipitate which dissolves to form a clear deep blue coloured liquid.

(c) Ammonia and acids

Ammonia reacts with nitric acid to produce ammonium nitrate:

$$NH_3 \text{ (g)} + HNO_3 \text{ (aq)} \rightarrow NH_4NO_3 \text{ (aq)}$$

Ammonia reacts with sulphuric acid to produce ammonium sulphate:

$$2NH_3 \text{ (g)} + H_2SO_4 \text{ (aq)} \rightarrow (NH_4)_2SO_4 \text{ (aq)}$$

(d) Ammonia and hydrogen chloride

Ammonia reacts with hydrogen chloride to produce fine particles of a white 'smoke' in the air. This is the test used to identify hydrogen chloride gas:

Ammonia + hydrogen chloride → ammonium chloride

NH_3 (g) + HCl (g) → NH_4Cl (s)

Ammonia is an important industrial chemical.

QUESTIONS

12 Compare the reactivity of ammonia with nitrogen.
13 How can ammonia be used to identify chemicals?

(iv) Oxygen

About 20% of the atmosphere is composed of oxygen. An oxygen molecule is formed from two oxygen atoms and is therefore a diatomic molecule. Oxygen is a colourless, odourless gas. It is not poisonous and produces a neutral response when tested with pH indicator paper. Oxygen is very reactive and combines with many other elements to form minerals in the Earth's crust. About 50% of the mass of the Earth's crust is due to oxygen atoms present in the minerals.

Oxygen does not burn but many other substances burn in it. The test for oxygen uses this property. If a gas is suspected of being oxygen a glowing splint is put into it. If oxygen is present the splint relights.

(a) The laboratory preparation of oxygen (Figure 9.8)

The source of oxygen is hydrogen peroxide. It releases oxygen very slowly into the air at room temperature. Manganese (IV) oxide is present as a catalyst. It speeds up the reaction at room temperature so enough oxygen is released to fill a test tube quickly.

The reaction for the production of oxygen is:

hydrogen peroxide → water + oxygen

$2H_2O_2$ (l) → H_2O (l) + O_2 (g)

Oxygen may also be produced by electrolysis.

It is produced industrially by fractional distillation of liquid air.

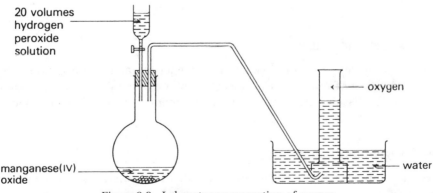

Figure 9.8 Laboratory preparation of oxygen

(b) Oxygen and life

Most of the life on the Earth is dependant on the presence of oxygen. It takes part in a chemical reaction called respiration in which energy is released from food. The energy that is released can then be used to provide the power for all the processes needed to keep an organism alive.

The equation for respiration is:

glucose + oxygen → carbon dioxide + water

$$C_6H_{12}O_6 \text{ (aq)} + 6O_2 \text{ (g)} \rightarrow 6CO_2 \text{ (g)} + 6H_2O \text{ (l)}$$

QUESTION

14 Compare the equation for respiration with equation for photosynthesis (see p. 122)

(c) Oxygen and burning

Burning is a process which needs three things – a fuel, heat and oxygen. They are usually represented as a triangle called the triangle of fire. If any side of the triangle is absent then burning will not occur. This is useful information for the fighting of fires. For example the heat may be removed by pouring water on a wood fire. The heat is taken up by the water and turns the water into steam. This also helps fight the fire by removing air from around the burning material.

However, water could not be used on a fire in which oil or petrol is the fuel as these substances float on the water. Water could also not be used on a burning piece of electrical equipment as water conducts electricity and a current may be conducted back up the water jet to the fire fighter. Oil, petrol and electrical fires are fought by removing oxygen. A chip pan fire can be covered with a damp towel and petrol on the road can be covered with sand.

An electrical fire can be fought by making sure the burning equipment is disconnected from the mains then using a carbon dioxide extinguisher.

QUESTION

15 Why do fire fighters remove trees in the path of a forest fire?

(d) Oxygen and water

Oxygen is only slightly soluble in water. Its presence in water supports a very wide range of life forms from the size of single cells to the whale shark and giant squid. When oxygen dissolves in water the pH of the water is not altered as it is a neutral gas.

(e) Oxygen and metals

Oxygen combines with metals to produce either **basic metal oxides** or amphoteric oxides.

Basic metal oxides

These oxides have an ionic structure and if they are soluble in water they dissolve to form alkaline solutions. Copper oxide and magnesium oxide are examples of basic metal oxides.

Amphoteric metal oxides

'Amphoteric' means having the properties of both metals and non-metals. These metal oxides have an ionic structure but they have acidic and basic properties. They can react with both acids and bases to form salts.

Examples of amphoteric oxides are zinc, aluminium and lead oxides.

These metals also have amphoteric hydroxides.

(f) Oxygen and non-metals

Oxygen combines with non-metals to form acidic or neutral compounds which have atoms bound together by covalent bonds. Carbon dioxide and sulphur dioxide are examples of acidic non-metal oxides and carbon monoxide and water are examples of neutral non-metal oxides.

QUESTION

16 Compare an amphoteric metal oxide with a non-metal oxide? Why can metal oxides be split into two groups?

(g) Uses of oxygen

Large amounts of oxygen are used in the production of steel. It is used to assist breathing in some hospital patients and for breathing purposes by divers, fire fighters, mountaineers and astronauts. It is carried by space rockets and used to burn fuel in the rocket's engines.

(v) Ozone

Ozone is an allotrope of oxygen. It is a blue gas with a strong smell. It is formed by the action of ultra violet light on oxygen molecules. The molecules are split into two oxygen atoms. Each atom then joins with an oxygen molecule to make an ozone molecule:

oxygen atom + oxygen molecule → ozone

$O (g) + O_2 (g) \rightarrow O_3 (g)$

Ozone forms a layer in the atmosphere 15–30 km above the Earth's surface. It prevents ultra violet light which is harmful to living things reaching the Earth's surface.

(vi) Sulphur

Sulphur is a pale yellow brittle solid. It has no smell. It escapes from inside the Earth when a volcano erupts and sublimes around the vent to form a

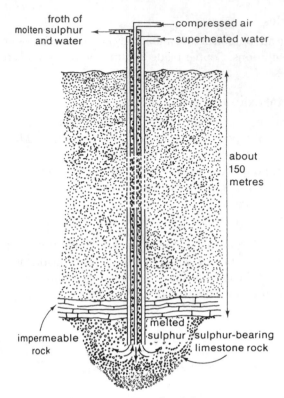

froth of
molten sulphur ← ← — compressed air
and water ← — superheated water

about
150
metres

impermeable
rock

melted
sulphur
sulphur-bearing
limestone rock

Figure 9.9 A Frasch sulphur pump

solid crust on the rocks. It is present in the fossil fuels – oil and natural gas – and is also found in large quantities in some limestone rocks where it forms sulphur beds.

Sulphur is extracted from crude petroleum oil and the sulphur beds that exist in the United States, Mexico and Poland. The sulphur is extracted from the beds using the Frasch sulphur pump (see Figure 9.9).

The sulphur is extracted from the sulphur bed in the following way. A pipe with two concentric tubes inside it is sunk to the sulphur bed. Super heated water at 170°C is pumped down to the sulphur bed where its heat melts the solid sulphur and turns it into a liquid. Hot compressed air is forced down the central tube and this pushes up the molten sulphur.

QUESTION

17 Describe the changes that take place in a sulphur bed when a Frasch sulphur pump is sunk into it and brought into operation.

(a) The allotropes of sulphur

When molten sulphur turns into a solid it forms crystals which look like tiny planks. They are called **monoclinic sulphur crystals**. If sulphur is allowed to crystallise from a solution which is at a temperature below 96°C tiny crystals in the shape of a rhombus called **rhombic sulphur** form.

(b) Sulphur and air

Sulphur burns in air or oxygen to form sulphur dioxide:

sulphur + oxygen → sulphur dioxide
$S (g) + O_2 (g) \rightarrow SO_2 (g)$

(c) Sulphur and solvents

Sulphur does not dissolve in water. It dissolves in some organic liquids such as methylbenzene which is used to prepare rhombic sulphur. Xylene is another example of an organic liquid in which sulphur dissolves.

(d) Uses of sulphur

Sulphur is a major raw material in the chemical industry. It is used to improve the natural properties of rubber, making it more elastic yet stronger. It is used to kill fungus that affects plants. Some sulphur compounds are used in the making of medicines.

(e) Sulphur dioxide

Sulphur dioxide is a colourless gas with a strong smell that makes people choke. If the gas is present in high concentrations, it can kill. It is an acidic compound.

(f) Sulphur dioxide and water

Sulphur dioxide reacts with water to produce sulphurous acid:

$SO_2 (g) + H_2O (l) \rightarrow H_2SO_3 (aq)$

When this reaction takes place in the atmosphere it contributes to the formation of acid rain.

(g) Sulphur dioxide as an oxidising agent

Sulphur dioxide reacts with hydrogen sulphide to produce sulphur and water:

$SO_2 (g) + 2H_2S (g) \rightarrow 3S (s) + 2H_2O (l)$

QUESTION

18 Identify the substances that have been reduced and oxidise in this reaction. (Use the information on p. 113 to help you.)

(h) Sulphur dioxide as a reducing agent

When sulphur dioxide is passed through a solution of potassium dichromate (VI) which contains a small amount of sulphuric acid the colour of the liquid changes from orange to green. This reaction is used to test for sulphur dioxide:

potassium dichromate (VI) + sulphur dioxide + sulphuric acid →
chromium (III) sulphate + potassium sulphate + water

$K_2Cr_2O_7$ (aq) + $3SO_2$ (g) + H_2SO_4 (aq) → $Cr_2(SO_4)_3$ (aq) + K_2SO_4 (aq) + H_2O (l)

(i) Uses of sulphur dioxide

Sulphur dioxide is used to disinfect food containers. It is used as the food preservative E220 and is used as a bleaching agent in the manufacture of textiles.

(vii) Chlorine

Chlorine is a halogen. It is a yellow green gas which has a strong smell and is poisonous. When tested with damp universal indicator paper it turns the paper red, then bleaches it. A molecule of chlorine is made from two atoms of chlorine. It is a diatomic molecule.

Chlorine can be made by heating concentrated hydrochloric acid with manganese (IV) oxide. If the acid is mixed with potassium manganate (VII) no heat is required. The following equation represents both reactions:

hydrochloric acid + oxidising agent → water + chlorine

$2HCl$ (aq) + [O] (aq or s) → H_2O (l) + Cl_2 (g)

(a) Chlorine and oxygen

Chlorine does not react with oxygen

(b) Chlorine and hydrogen

When hydrogen gas is burnt in chlorine hydrogen chloride is produced:

hydrogen + chlorine → hydrogen chloride

H_2 (g) + Cl_2 (g) → $2HCl$ (g)

(see hydrogen chloride, p. 133).

(c) Chlorine and water

Chlorine is soluble in water.

It forms chlorine water which is a mixture of hydrochloric and hypochlorous acid. The hypochlorous acid slowly decomposes to hydrochloric acid and oxygen. The dissolving of chlorine in water can be summarised by the following equation:

chlorine + water → hydrochloric acid + oxygen

$2Cl_2$ (g) + $2H_2O$ (l) → $4HCl$ (aq) + O_2 (g)

(d) Chlorine and sodium hydroxide solution

Chlorine reacts with cold sodium hydroxide solution to produce sodium chloride and sodium chlorate (I):

Cl_2 (g) + $2NaOH$ (aq) → $NaCl$ (aq) + $NaOCl$ (aq) + H_2O (l)

(e) Chlorine and metals

When chlorine is heated with iron it forms iron (III) chloride:

$$3Cl_2 \text{ (g)} + 2Fe \text{ (s)} \rightarrow 2FeCl_3 \text{ (s)}$$

When chlorine is passed into solutions of iron (II) compounds it oxidises them to iron (III) compounds:

$$Cl_2 \text{ (g)} + 2Fe^{2+} \text{ (aq)} \rightarrow 2Cl^- \text{ (aq)} + 2Fe^{3+} \text{ (aq)}$$

(f) Uses of chlorine

Chlorine is used in the manufacture of bleaches and disinfectants. It is the substance which kills bacteria. Chlorine is also used in the treatment of water to make it fit to drink.

Chlorine can cause an explosion if it comes into contact with ammonia. Therefore chlorine containing bleaches should be kept away from ammonia-containing cleaning products.

(viii) Hydrogen chloride

Hydrogen chloride has a strong smell. It is a colourless gas in dry air but produces visible fumes in normal moist air. If hydrogen chloride is tested with damp indicator paper, it turns the paper red.

The test for the gas is to pass it into silver nitrate solution where it will form a white precipitate of silver chloride. Hydrogen chloride also forms a white smoke with ammonia.

(a) Preparation of hydrogen chloride

Hydrogen chloride can be produced by adding concentrated sulphuric acid to sodium chloride, as the equation shows:

sulphuric acid + sodium chloride → sodium hydrogen sulphate
+ hydrogen chloride

$$H_2SO_4 \text{ (l)} + NaCl \text{ (s)} \rightarrow NaHSO_4 \text{ (s)} + HCl \text{ (g)}$$

(b) Hydrogen chloride and water

Hydrogen chloride dissolves readily in water and forms hydrochloric acid (see also p. 151):

$$HCl \text{ (g)} + H_2O \text{ (l)} \rightarrow H_3O^+ \text{ (aq)} + Cl^- \text{ (aq)}$$

9.3 Comparing the reactivity of the non-metals

The reactivity of non-metals can be compared by displacement reactions. These reactions for the halogens are considered on p. 100. Further to the results described there, it is found that chlorine displaces oxygen from hydrogen in water (see p. 132) but that the other halogens to not produce such a displacement. This sets chlorine above oxygen in a table of reactivity.

If the three halogens are each added to a separate solution of sodium sulphide a white precipitate is formed in all three liquids. The precipitate is formed as the sulphur is displaced from the compound. The following equation describes the reaction that occurs between bromine and sodium sulphide:

bromine + sodium sulphide → sodium bromide

Br_2 (aq) + Na_2S (aq) → S (s) + 2NaBr (aq)

This reaction shows that all three halogens are more reactive than sulphur

QUESTIONS

19 Construct a symbol equation for the reaction between chlorine (Cl_2) and sodium sulphide (Na_2S).
20 Construct a reactivity series for the non-metals bromine, sulphur, iodine, oxygen, chlorine, starting with the most reactive element.

9.4 Summary

- The physical properties of non-metals include low melting and boiling points and low densities. Almost all do not conduct electricity (see p. 118).
- Non-metals do not react with dilute acids (see p. 119).
- Carbon takes part in a range of reactions and has a variety of uses (see p. 119).
- Carbon dioxide has a wide range of uses (see p. 122).
- Carbon moves through the environment in the carbon cycle (see p. 123).
- Nitrogen is an unreactive gas (see p. 123).
- Nitrogen moves through the environment in the nitrogen cycle (see p. 124).
- Ammonia is a gas which is soluble in water and reacts with acids and hydrogen chloride (see p. 126).
- Oxygen is a reactive gas which supports life and burning (see p. 127).
- Sulphur is extracted by the Frasch process (see p. 130).
- Sulphur has two allotropes, reacts with oxygen and dissolves in some organic liquids (see p. 130).
- Sulphur dioxide is both an oxidising agent and a reducing agent (see p. 131).
- Chlorine reacts with hydrogen, water, sodium hydroxide and iron. It also kills bacteria (see p. 132).
- Hydrogen chloride fumes in moist air and dissolves in water to form hydrochloric acid (see p. 133).
- Some non-metals can displace each other and be arranged in a reactivity series (see p. 133).

▪ ⋁ **10** Aqueous chemistry

Objectives

When you have competed this chapter you should be able to:
- Understand the **water cycle**
- Explain how water is prepared for **drinking**
- Understand how substances **dissolve in water**
- Recognise the **salts** that are soluble and insoluble in water
- Understand how **cations** and **anions** can be detected
- Describe how **stalactites** and **stalagmites** are formed
- Explain the test for finding the **hardness of water**
- Evaluate the advantages and disadvantages of **hard water**
- Describe how hardness can be **removed from water**
- Explain how chemicals may be **extracted from sea water**.

From space the Earth looks blue. This is due to about 80% of its surface being covered by water. It has been estimated that there are 1.5 million, million, million litres of water on the planet and over 95% of it is in the seas and oceans. The rest is in the form of snow and ice, streams, rivers, lakes or in liquid form beneath the ground surface or vapour form in the atmosphere.

The way that water circulates through the atmosphere is shown in Figure 10.1.

QUESTION

1 Describe the path water in a cloud may take after falling as rain until it enters another cloud. Include the processes which take place during the path.

10.1 Water for drinking

Over 70% of your body is composed of water but it is constantly escaping in the processes of sweating, respiring, urinating and defecating. It must be replaced daily to remain healthy and you could probably only survive for about six days without it.

Most drinking water is collected from streams and rivers and stored in reservoirs. Some water is collected from water which has collected underground.

At the water works the water first passes through a grid which removes large items such as twigs or rubbish such as plastic bags (Figure 10.2). The water is then

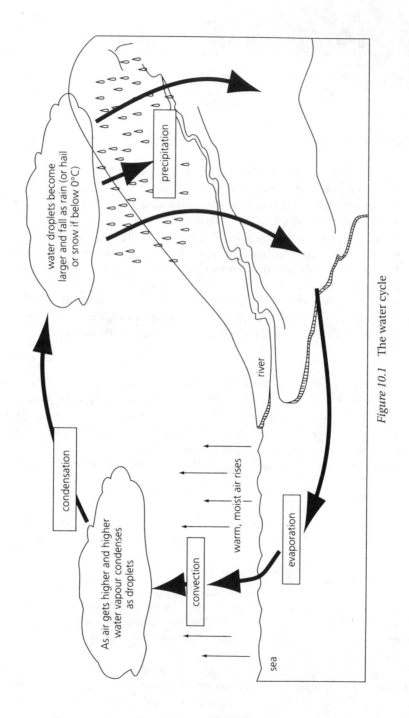

Figure 10.1 The water cycle

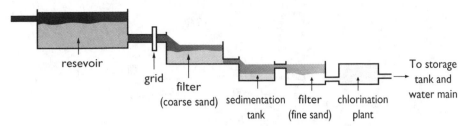

Figure 10.2 The path of water through a typical water works

filtered through coarse sand which also contains a film of microorganisms. The coarse sand traps the larger particles in the water and microorganisms feed on some of the bacteria in the water. When the water emerges from the coarse sand filter it may still have tiny particles such as clay suspended in it. These particles are removed in a process called **sedimentation**. Chemicals called **flocculants** such as potassium aluminium sulphate which is also called potash alum are poured into the water. They cause the tiny particles to stick together and form larger particles which are pulled down by the force of gravity to the floor of the sedimentation tank. The water is filtered again, but this time it passes through finer sand to remove all the tiny solid particles remaining, then it is treated with chlorine so that any remaining bacteria can be destroyed.

QUESTION

2 What may be found in the water taken in at a water works and what process are used to remove them?

10.2 The water molecule

Water is made from one atom of oxygen and two atoms of hydrogen, as Figure 10.3 shows.

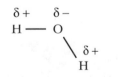

Figure 10.3 A water molecule

The single electron which is part of each hydrogen atom is shared with the oxygen atom. This makes the hydrogen atoms slightly electropositive and the oxygen atom slightly electronegative. These small charges on the parts of the oxygen molecule cause the hydrogen atoms of one molecule to be attracted to the oxygen atoms of other molecules. This force of attraction is called a

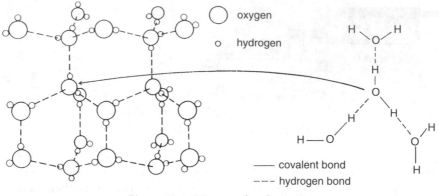

Figure 10.4   Water molecules in ice

hydrogen bond. It is weaker than the covalent bond that exists between the atoms in the water molecules but is strong enough to affect the melting and boiling points of water and make them higher than would be the case if they did not exist. The hydrogen bonds also make the water molecules take up the form shown in Figure 10.4 when ice is formed instead of the molecules becoming packed together as occurs in other substances.

This open structure which forms ice is less dense than the liquid water from which it formed so the ice floats at the water surface. The hydrogen bonds also help water to dissolve ionic substances. Most covalent molecules do not dissolve in water but some such as sugar do dissolve because they have regions which are electronegative and electropositive like water molecules.

10.3 Solubility

- A substance which dissolves in another is called a **solute**
- A substance which dissolves another is called a **solvent**
- A solute and solvent together make a mixture called a **solution**.

(i) Gases

(a) Gases and temperature

Many gases dissolve in water. Their solubility in water increases as the temperature of the water decreases. For example, cold water holds more dissolved oxygen than warm water. If warm water from a power station is released into a river the temperature of the river water increases and the solubility of the oxygen decreases. This threatens the lives of aquatic animals which take oxygen from water when they breathe. Some life forms require such high levels of oxygen in the water that they are only found in mountain streams where the water is permanently cold.

(b) Gases and pressure

Gases can be compressed, and when their pressure is increased and they are presented to a liquid the solubility of the gas in the liquid increases. Carbon dioxide is used to give a refreshing taste to a fizzy drink. High pressure is used to make it dissolve in large amounts. When the pressure on the drink is reduced, as a can or a bottle is opened, large amounts of the gas escape forming bubbles and make the drink fizz.

QUESTION

3 What factors affect the solubility of a gas in a liquid?

(ii) Solids

Ionic substances and a few covalent substances dissolve in water. Organic liquids are better solvents of covalent compounds than water.

The speed at which a solid solute can be made to dissolve can be increased in four ways:

(1) **The surface area of the solute can be increased.** For example, granulated sugar dissolves faster than a sugar lump.
(2) The **solvent can be stirred** as the solute is added.
(3) The **volume of the solvent** can be increased to provide more space for the solute.
(4) The **temperature** of the solvent may be **increased**.

QUESTION

4 What are the factors that affect the speed of dissolving a solid in a liquid?

10.4 Solubility and temperature

A solute may be added to a solvent until the solvent can dissolve no more. When this happens the solution that is produced is called a **saturated solution**. The solubility of a solute is described as the amount of it in grams which dissolves in 100 g of the solvent.

The solubility of a solute in a solvent depends on the temperature. For most solutes their solubility increases as the temperature rises. The exact way the solubility changes can be discovered by measuring how many grams of solute dissolve in 100 g of solvent at a particular temperature, then changing the temperature of another 100 g of the solvent and finding how many grams of solute

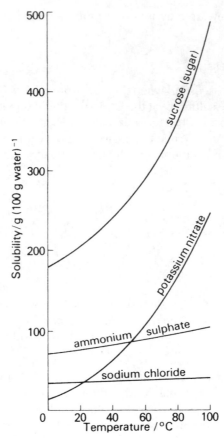

Figure 10.5 Solubility of four solutes in water

then dissolve. By repeating this process over a range of temperatures and plotting the solubility of the solute against the temperature of the solvent a solubility curve can be produced.

Figure 10.5 shows the solubility of four solutes in water.

QUESTION

5   How does the solubility of the four solutes compare over the temperature range of the solvent of 0°C to 60°C?

10.5 The solubility of different salts

When a substance does not dissolve in a solvent it is said to be **insoluble**.

The solubility of salts in water varies. Some salts are insoluble. Table 10.1 shows the different salts which are soluble and insoluble in water.

Salt	Soluble	Insoluble
ammonium salts	all	none
carbonates	ammonium potassium sodium	the rest
chlorides	all except	lead, silver
nitrates	all	none
potassium	all	none
sodium	all	none
sulphates	all except	calcium barium, lead

Table 10.1 Soluble and Insoluble salts

QUESTIONS

6 Which salts are the least soluble in water?
7 Which of the following are (a) soluble, (b) insoluble in water? Sodium carbonate, magnesium carbonate, aluminium chloride, silver chloride, ammonium nitrate, lead nitrate, lead sulphate, aluminium sulphate.

10.6 Testing for ions

The ions in a salt separate when the salt is dissolved in water. Positively charged ions are called **cations** and negatively charged ions are called **anions**.

(i) Testing for cations

The common cations dissolved in water can be identified by dividing the water sample into two and testing one part with sodium hydroxide and the other part with ammonia solution.

(a) Testing with sodium hydroxide solution

Table 10.2 shows what happens in the solution after sodium hydroxide has been added if one of the common cations is present.

QUESTIONS

8 Three cations can be identified with sodium hydroxide solution. Which are they?
9 Which cations does the test with sodium hydroxide fail to identify?

(b) Testing with ammonia solution

Table 10.3 shows what happens in the solution after ammonia solution has been added if some of the common metal cations are present.

Cation	Change in solution	Change when excess added
potassium	none (see flame test)	none
sodium	none (see flame test)	none
calcium	white precipitate	insoluble
magnesium	white precipitate	insoluble
aluminium	white precipitate	soluble
zinc	white precipitate	soluble
iron (II)	greeny blue precipitate	insoluble
iron (III)	brown precipitate	insoluble
lead	white precipitate	soluble
copper	pale blue precipitate	insoluble
ammonium	none (see separate test)	none

Table 10.2 Change in solution when cation present

Cation	Change in solution	Change when excess added
calcium	none	none
aluminium	white precipitate	insoluble
zinc	white precipitate	soluble
iron (II)	greeny blue precipitate	insoluble
iron (III)	brown precipitate	insoluble
lead	white precipitate	insoluble
copper	blue precipitate	blue solution

Table 10.3 Change in solution when common metal cations present

QUESTIONS

10 Which three metal ions are soluble in excess ammonia solution?

11 How can the results of the test shown in Tables 10.2 and 10.3 be used to distinguish between calcium and aluminium ions?

12 How do lead ions react differently with sodium hydroxide solution and ammonia solution?

(c) Testing for the ammonium ion

If the solution has failed to change when sodium hydroxide is added it may contain ammonium ions. To confirm this, the mixture should be gently heated. If ammonium ions are present ammonia gas will be released.

QUESTION

13 What cations may be present if ammonia is not produced when the mixture is gently heated?

(d) Flame test (Table 10.4)

This test can be used to identify the presence of some metal ions.

A piece of new nichrome wire is needed for the test. It is dipped in concentrated hydrochloric acid and then in a sample of the solid salt. When the acid meets the salt a chloride of the metal forms on the wire. If the end of the wire is then put into the flame of a Bunsen burner a coloured flame may be seen rising off the end of the wire.

Metal	Colour of flame
potassium	lilac
sodium	yellow
calcium	red
magnesium	none
copper	greeny blue

Table 10.4   Flames produced by metals

QUESTION

14   A solution containing cations does not change when sodium hydroxide solution is added to it. What test would have to be carried out to show that the only cations it contained were sodium ions?

(ii) Testing for anions

If a salt solution fizzes when dilute nitric acid is added to it the gas should be tested with lime water. If the lime water turns cloudy the gas is carbon dioxide and the ion present is a carbonate ion.

If the salt solution does not fizz with dilute nitric acid it may contain other ions and the acidified solution can be tested with the following reagents:

(1) *Silver nitrate*
    If chloride ions are present a white precipitate is produced.
    If iodide ions are present a yellow precipitate is produced.
(2) *Barium chloride*
    If sulphate ions are present a white precipitate is produced.

If these reagents fail to produce a precipitate the following test should be tried on a sample of the solid salt. Sodium hydroxide and aluminium powder should be added to the solid salt. If ammonia is formed, the salt contains nitrate ions.

10.7 Reactions of metals with water

(See reactivity series, p. 108.)

10.8 Reactions of non-metals with water

(See Chapter 9 for reactions with carbon, nitrogen, oxygen, sulphur and chlorine.)

10.9 Stalactites and stalagmites

Figure 10.6 shows a cave in limestone rock. Stalactites hang from the roof and stalagmites rise from the floor. They have formed because of the solvent properties of water.

Some of the carbon dioxide in the air dissolves in the water droplets in clouds. A solution of carbonic acid forms. This reaction is reversible (see p. 75). The raindrops carry the carbonic acid to the ground. If the rain water passes into the ground made by limestone rock it reacts with the calcium carbonate present to form calcium hydrogen carbonate which dissolves slightly in the water. If the rain water reaches the roof of a cave in the rock it collects then drips onto the cave floor. At the place on the roof where the drops form, some of the carbon dioxide leaves the water and re-enters the atmosphere. Particles of calcium carbonate form which stick to the cave roof. In time, as the water drips from the same place, a pile of particles collects there and begins to form a stalactite.

As the drop falls and splashes on the floor more carbon dioxide leaves the water and particles of calcium carbonate collect where the water makes contact

Figure 10.6 Stalactites and stalagmites

with the ground. In time, a pile of particles collects and there begins to form a stalagmite.

QUESTION

15 How do you think the columns of rock in the cave in Figure 10.6 formed?

10.10 Hardness of water

The hardness of water is due to the presence of some soluble calcium and magnesium salts. These salts are:

(1) *Calcium salts*
  • Calcium hydrogen carbonate
  • Calcium sulphate
  • Calcium chloride
(2) *Magnesium salts*
  • Magnesium hydrogen carbonate
  • Magnesium sulphate
  • Magnesium chloride

The calcium and magnesium ions formed when these salts dissolve are responsible for the hardness of water. There are two kinds of hardness of water – temporary hardness and permanent hardness. A sample of water may have both kinds of hardness.

(i) Temporary hardness

This type of hardness can be removed by heating the water. It is caused by the hydrogen carbonate salts of calcium and magnesium. These salts decompose on heating to produce the carbonate salts of the metal. The carbonates are insoluble compounds in water and when they form, the calcium and magnesium ions are removed from the solution:

Calcium hydrogen carbonate → calcium carbonate + carbon dioxide + water

$$Ca(HCO_3)_2 \text{ (aq)} \rightarrow CaCO_3 \text{ (s)} + CO_2 \text{ (g)} + H_2O \text{ (l)}$$

The calcium carbonate forms a precipitate which settles out on the inside of kettles, where it forms furring. On the inside of boilers it forms a deposit called **scale**.

These deposits can be removed by using weak acids such as ethanoic acid (vinegar) and citric acid. The use of stronger acids could cause damage to the metal beneath the deposits.

QUESTION

16 Construct a symbol equation to show how magnesium hydrogen carbonate $Mg(HCO_3)_2$ in water decomposes when the water is heated.

(ii) Permanent hardness

This type of hardness is caused by calcium sulphate and chloride and magnesium sulphate and chloride. These compounds do not decompose when they are heated.

(a) Testing the hardness of water

Hard water does not cause a soap to produce lather easily. Before a lather can be produced the soap has to remove the calcium and magnesium ions. This is done by the formation of calcium stearate which forms a precipitate known as scum.

sodium stearate + calcium hydrogen carbonate
(soap)
$\rightarrow$ calcium stearate + sodium hydrogen carbonate
(scum)

In the test for hardness, soap solution is added in volumes of $0.5\,cm^3$ from a burette until it can produce a lather which lasts for a minute after the soapy water has been shaken for 10 seconds. The hardness of the water is assessed by the amount of soap which has to be added to produce the one-minute lather.

QUESTION

17 A sample of water needed $3\,cm^3$ of soap solution to produce a lather. When a second sample of the same water was heated then tested, only $2\,cm^3$ of water was needed. Explain this result.

(b) The advantages of hard water

The calcium ions in the water are used by the body for the formation of bones and teeth. It is believed that drinking hard water reduces the development of heart disease. Hard water has a pleasant taste. It is a good raw material for the brewing of beer.

(c) The disadvantages of hard water

Large amounts of soap are required for washing and the deposits of kettles and boilers absorb energy when the water is heated. This makes the kettles and boilers less efficient and more energy is needed to bring the water to the required temperatures.

QUESTION

18 What may happen in a hot water pipe in a hard water area that is used constantly for a long time?

(d) Removing hardness in water

(i) Using washing soda

Washing soda is sodium carbonate. When it is added to water containing soluble calcium and magnesium salts insoluble calcium and magnesium carbonates form:

Magnesium chloride + sodium carbonate → magnesium carbonate + sodium chloride

$MgCl_2$ (aq) + Na_2CO_3 (aq) → $MgCO_3$ (s) + 2NaCl (aq)

QUESTION

19 Produce a symbol equation to show the reaction between calcium sulphate ($CaSO_4$) and sodium carbonate (Na_2CO_3).

(ii) Using a water softener

A water softener contains a substance called an ion exchange resin. It contains sodium ions. When hard water is run through the water softener the calcium ions and magnesium ions are taken up by the resin and the sodium ions are released. The water leaving the resin is called **soft water** and forms a lather very easily with soap. When all the sodium ions have been released the water softener is recharged by passing salt solution through it. This supplies sodium ions to the resin and removes the calcium ions.

(iii) Distillation

Water is heated until it boils and the steam is collected and cooled to form very pure water. The substances dissolved in the water remain in the container where the water is boiled.

10.11 Chemicals from sea water

(i) Sea salt

Figure 10.7 shows a group of salt pans. They are set up on the coasts in the tropical regions of the world where the air temperature is high. Sea water is run into a salt pan and left for evaporation to take place. As more water evaporates from the salt pan the solution left behind becomes more concentrated until it becomes saturated. When this occurs the salt begins to crystallise and it is removed. Sea salt is mainly composed of sodium chloride but it also contains a range of other salts such as magnesium chloride.

(ii) Bromine

The first stage in the extraction of bromine is similar to that of extracting salt. The sea water is allowed to evaporate to make the water more concentrated. The sea water is then acidified with sulphuric acid to make a solution with a pH of 3.5. Chlorine gas is then bubbled through the warm acidic solution and the bromine is displaced as the equation shows:

chlorine + bromide ions → bromine + chloride ions

Cl_2 (g) + 2Br^- (aq) → Br_2 (aq) + 2Cl^- (aq)

Figure 10.7 Extraction of salt from sea water

Steam is blown through the solution to release the bromine as a vapour. The steam and bromine are allowed to condense and the bromine is purified by distillation.

QUESTION

20 How does using the Sun's energy affect the cost of extracting the chemicals? Explain your answer.

10.12 Summary

- Water passes through the environment in the water cycle (see p. 136).
- Water is treated at a water works to make it fit to drink (see p. 135).
- The solubility of gases in water is affected by pressure and temperature (see p. 138).
- The solubility of a solid in water is affected by the temperature (see p. 139).
- Salts may be soluble or insoluble in water (see p. 140).
- There are tests to identify cations dissolved in water (see p. 141).
- The flame test is used to identify some metal ions (see p. 143).
- There are tests to identify anions dissolved in water (see p. 143).
- Stalactites and stalagmites form in limestone caves (see p. 144).
- Some calcium and magnesium salts cause hardness in water (see p. 145).

- Soap solution is used to measure hardness in water (see p. 146).
- There are advantages and disadvantages in using hard of water (see p. 146).
- Hard water may be made into soft water (see p. 146).
- Some chemicals are extracted from sea water (see p. 147).

▣ Ⴙ 11 Acids, bases and salts

Objectives

When you have completed this chapter you should be able to:
- Distinguish between **mineral** and **organic acids**
- Explain the difference between **strong** and **weak acids**
- Describe **bases**
- Understand the **neutralisation reaction**
- Distinguish between a **base** and an **alkali**
- Explain the difference between **stong** and **weak alkalis**
- Describe how **soluble salts** are prepared
- Describe how **insoluble salts** are prepared by precipitation
- Describe how a salt can be prepared by a **synthesis reaction**.

The word acid comes from a Latin word meaning sour. You may safely taste the sourness of an acid if you suck a slice of lemon but NO ACIDS OR ANY OTHER CHEMICALS SHOULD BE TASTED IN THE LABORATORY.

11.1 The two kinds of acids

There are two kinds of acids – **organic acids** and **mineral acids**.

Examples of organic acids are tartaric acid found in grapes and ethanoic acid found in vinegar. The three most common mineral acids are hydrochloric acid, sulphuric acid and nitric acid acids can be corrosive (Figure 11.1).

11.2 Simple tests for acids

An acid will turn blue litmus paper red. It will produce hydrogen gas if it is tested with a metal from the middle of the reactivity series such as magnesium or iron. An acid will also produce carbon dioxide gas if it is added to a carbonate or a hydrogen carbonate.

QUESTION

1 Construct word and symbol equations for the reaction between hydrochloric acid (HCl) and (a) iron (Fe), (b) calcium carbonate ($CaCO_3$).

CORROSIVE

Figure 11.1 Corrosive symbol used for acids and bases

11.3 The essential part of an acid

An acid is a substance which dissolves in water and releases hydrogen ions.

For example when hydrogen chloride dissolves in water the following ions are produced:

$$HCl\ (g) + (aq) \rightarrow H^+\ (aq) + Cl^-\ (aq)$$

In this equation, the H^+ ion is called the **hydrated hydrogen ion**.

Another way of writing the equation is:

$$HCl\ (g) + H_2O\ (l) \rightarrow H_3O + (aq) + CL^-\ (aq)$$

In this equation, the H_3O^+ ion is called the **hydroxonium ion**.

It is the presence of hydrogen ions which give an acid its properties.

QUESTIONS

2 What are the symbols for the hydrated hydrogen ion and the hydroxonium ion?
3 Nitric acid has the formula HNO_3. Show two ways of displaying its ions in equations.

11.4 Strong and weak acids

In hydrochloric acid and the other mineral acids all the ions separate and remain separated. In ethanoic acid only some of the ions separate and a reversible reaction occurs which causes some of the ions to recombine:

$$CH_3CO_2H\ (l) + H_2O\ (l) \rightleftharpoons H_3O+\ (aq) + CH_3CO^{2-}\ (aq)$$

There are far fewer hydrogen ions in a volume of ethanoic acid than there are in an equal volume of hydrochloric acid that has the same concentration. The complete ionisation of the hydrochloric acid characterises it as a strong acid. The much smaller degree of ionisation of the ethanoic acid characterises it as a weak acid. It is important to realise that the strength of the acid depends initially on its degree of **ionisation**, and not its concentration, although it is possible to have a concentrated weak acid with the same number of hydrogen ions in it as a very dilute strong acid.

The amount of hydrogen ions present in an acid is measured on the **pH scale**. Universal indicator paper is used to identify the pH of an acid. The pH is less than 7. The weakest acid has a pH of 6 and the strongest acid has a pH of 0. A precise record of the pH of an acid can be made by dipping the probe of a pH meter into the acid.

QUESTION

4   How are strong and weak acids different?

11.5 Bases

A base is a substance which can neutralise an acid. When a base reacts with an acid a salt and water are formed. The compounds which are bases are metal oxides, metal hydroxides, and metal carbonates. There are two kinds of bases – bases which are insoluble in water and bases which are soluble in water.

(i) Insoluble bases

Most bases are insoluble in water. The exceptions are listed in the next section. A solid substance is tested to see if it is a base in the following way. Universal indicator solution is added to a sample of acid and a red colour is observed. Small pieces of the solid are then added to the acid and the colour of the indicator is observed. If the solid is a base the colour of the indicator should turn orange and then yellow showing that the pH of the solution is rising as the acid is being neutralised. This process is complete when the indicator turns green.

(ii) Soluble bases

Soluble bases are called **alkalis**. This word is derived from an Arabic word which means plant ash because the carbonates of potassium and sodium were obtained by burning plants.

The alkalis used in the laboratory are:

• Potassium hydroxide solution
• Sodium hydroxide solution

- Calcium hydroxide solution
- Barium hydroxide solution
- Ammonia solution.

11.6 Simple test for an alkali

An alkali turns red litmus blue.

QUESTION

5 How is an acid different from a base?

11.7 Strong and weak alkalis

When a base dissolves in water to form an alkali, **hydroxide ions** are released. They may be released as the substance dissolves. For example, when sodium hydroxide dissolves in water the following ions are produced:

sodium hydroxide + water → sodium ions + hydroxide ions
$NaOH (s) + (aq) \rightarrow Na^+ (aq) + OH^- (aq)$

The hydroxide ions may form as a consequence of the base reacting with the water and not simply dissolving in it. For example, calcium oxide reacts with water:

calcium oxide + water → calcium ions + hydroxide ions
$CaO (s) + H_2O (l) \rightarrow Ca^{2+} (aq) + 2OH^- (aq)$

In sodium hydroxide solution, all the hydroxide ions separate and remain apart. In ammonia solution only some of the ions separate and a reversible reaction occurs which causes some of the ions to recombine:

ammonia + water $\rightleftharpoons$ ammonium ions + hydroxide ions
$NH_3 (g) + H_2O (l) \rightleftharpoons NH_{4+} (aq) + CH_3CO^{2-} (aq)$

There are far fewer hydroxide ions in a volume of ammonia solution than there is in an equal volume of sodium hydroxide solution that has the same concentration. The complete ionisation of the sodium hydroxide solution characterises it as a strong alkali. The much smaller degree of ionisation of the ammonia solution characterises it as a weak alkali. It is important to realise that the strength of the alkali depends initially on its degree of **ionisation** and not its concentration, as in the case of acids.

The amount of hydroxide ions present in an alkali is measured on the pH scale. Universal indicator paper is used to identify the pH of an alkali. The pH is more than 7. The weakest alkali has a PH of 8 and the strongest alkali has a pH of 14. A precise record of the pH of an alkali can be made by using a pH meter.

QUESTION

6 How are acids and alkalis (a) different, (b) similar?

11.8 The preparation of salts

A salt is a compound which has a positive ion formed from a metal or an ammonium ion and a negative ion formed from a non-metal or an acid. The salts produced by the three most common mineral acids are chlorides from hydrochloric acid, nitrates from nitric acid and sulphates from sulphuric acid.

11.9 Soluble salts

(i) Solid and acid

In all the three following methods, the solid which is in powder form is added to excess so that the acid is neutralised. This reaction produces water in which the excess solid does not dissolve.

(a) Metal and acid

The metals which are used in this method are those in the middle of the reactivity series. They are magnesium, aluminium, zinc, iron and tin. The following equation shows a typical reaction:

magnesium + sulphuric acid → magnesium sulphate + hydrogen
$Mg (s) + H_2 SO_4 (aq) → MgSO_4 (aq) + H_2 (g)$

As the reaction takes place, effervescence (fizzing) occurs. This is due to the escape of hydrogen gas. When neutralisation occurs the hydrogen is no longer released.

QUESTIONS

7 How can you tell that neutralisation has occurred when powdered metal is added to an acid?
8 Construct symbol equations for the making of (a) magnesium chloride ($MgCl_2$), (b) zinc sulphate ($ZnSO_4$).

(b) Metal oxide (base) and acid

Metal oxides which are used in this method range from magnesium to copper in the reactivity series. The following equation shows a typical reaction:

magnesium oxide + sulphuric acid → magnesium sulphate + water

$$MgO \text{ (s)} + H_2SO_4 \text{ (aq)} \rightarrow MgSO_4 \text{ (aq)} + H_2O \text{ (l)}$$

QUESTIONS

9 How can you tell when neutralisation has occurred in a reaction between a base and an acid? (*Hint:* A base is insoluble in water.)

10 Construct symbol equations for the making of (a) magnesium chloride ($MgCl_2$) from magnesium oxide (MgO), (b) zinc sulphate ($ZnSO_4$) from zinc oxide (ZnO).

(c) Metal carbonates and acid

All metal carbonates can be used in this method. The following equation shows a typical reaction:

Magnesium carbonate + sulphuric acid → magnesium sulphate + water + carbon dioxide

$$MgCO_3 \text{ (s)} + H_2SO_4 \text{ (aq)} \rightarrow MgSO_4 \text{ (aq)} + H_2O \text{ (l)} + CO_2 \text{ (g)}$$

As the reaction takes place, effervescence occurs. This is due to the production of carbon dioxide gas. When neutralisation occurs the effervescence stops.

QUESTIONS

11 How can you tell when neutralisation has occurred when a metal carbonate has been added to an acid?

12 Write the symbol equations for the production of (a) magnesium chloride ($MgCl_2$) from magnesium carbonate ($MgCO_3$), (b) sodium chloride ($NaCl$) from sodium carbonate (Na_2CO_3).

(ii) Solution and acid

(a) Alkali and acid

This method is used to prepare potassium, sodium and ammonium salts. The titration method is used to neutralise the acid (see p. 169). An indicator is used at first to find when neutralisation occurs. The procedure is then repeated without the indicator so that a pure salt may be produced. The following equation shows a typical reaction:

sodium hydroxide + hydrochloric acid → sodium chloride + water
$$NaOH \text{ (aq)} + HCl \text{ (aq)} \rightarrow NaCl \text{ (aq)} + H_2O \text{ (l)}$$

QUESTIONS

13 Construct symbol equations for the production of (a) potassium sulphate (K_2SO_4); and (b) sodium nitrate ($NaNO_3$).

14 What are the different ways in which neutralisation can be detected in the preparation of salts?

(iii) Preparing crystals from a solution of soluble salts

The salts produced in the previous four methods are all soluble. When neutralisation occurs excess reactant remains undissolved. This solid is removed by filtration. The water is removed by evaporation. This is done in two stages. First, the solution is poured into an evaporating dish and heated.

As the water evaporates the concentration of the salt in the solution increases until the solution becomes a saturated solution. The test to discover when the solution is saturated is carried out as follows. The end of a glass rod is dipped into the solution then removed. The end, now covered in liquid, is allowed to cool. If crystals form as the rod cools the solution is saturated.

When the solution has become saturated the heat is removed from the evaporating dish and the solution is allowed to cool. Evaporation continues more slowly and crystals of the salt form in the dish. At this stage, the solution and the crystals can be separated by filtration. The crystals can then be cleaned by placing them on filter paper and washing them with a small amount of distilled water. The wet crystals should then be dried with tissue paper.

(iv) Water of crystallisation

The crystals of most salts must not be heated to complete the drying process because they are **hydrated crystals**. This means that they contain water of crystallisation which is water that is bonded to the compound and gives the crystal its shape and colour. For example, the formula of hydrated copper sulphate which forms blue crystals is $CuSO_4$ $5H_2O$.

If a hydrated crystal is heated, a **dehydration reaction** takes place in which the water escapes as vapour. When hydrated copper sulphate crystals are heated a white powder is left behind as the water escapes. The symbol equation for this change is:

$$CuSO_4 \cdot 5H_2O \text{ (s)} \rightarrow CuSO_4 \text{ (s)} + 5H_2O \text{ (l)}$$

blue   grey-white

Hydrated crystals can be dried by leaving them in a warm place.

Crystals which are not hydrated, like those of sodium chloride, can be gently heated in the evaporating dish to prevent the product spitting until all the water is removed.

QUESTION

15 What are the stages in preparing crystals of a metal salt from (a) a metal and acid, (b) an alkali and acid?

11.10 Insoluble salts

Insoluble salts can be prepared by a precipitation reaction. It is sometimes called a double decomposition reaction because both reactants decompose to form the products

A general word equation for the reaction is:

soluble salt + soluble salt → insoluble salt + soluble salt

One soluble salt reactant has the metal ions that are required and the other soluble salt reactant has the non-metallic ions that are required.

A general formula equation for the reaction is:

MNx (aq) MxN (aq) → MN (s) + MxNx (aq)

where M = metal ion required

Mx = metal ion not required

N = non-metal ion required

Nx = non-metal ion not required.

A precipitation reaction occurs when barium chloride is added to sodium sulphate:

$BaCl_2$ (aq) + Na_2SO_4 (aq) → $BaSO_4$ (s) + $2NaCl$ (aq)

Although salts are often called insoluble, it would be more precise to say that they are relatively insoluble as they are all sparingly soluble in water.

The two reactants are stirred together to ensure they are completely mixed. The insoluble salt forms at once. It can be removed from the mixture by filtration, washed to remove any remaining soluble product then dried in air or a warm oven.

11.11 Direct combination

Some salts can be prepared by direct combination of the metal and the non-metal. For example when iron is heated strongly with sulphur the salt iron (II) sulphide is produced:

Fe (s) + S (s) → Fe(II)S (s)

This is an example of a synthesis reaction.

11.12 Summary

- There are two kinds of acids – mineral acids and organic acids (see p. 150).
- An acid produces hydrogen ions in water (see p. 151).
- A base is a substance which can neutralise an acid (see p. 152).
- Soluble salts can be prepared from acids and metals, metal oxides, metal carbonates and alkalis (see p. 154).

- Crystals of salts can be prepared by careful evaporation (see p. 156).
- Insoluble salts are prepared by precipitation (see p. 157).
- Some salts can be prepared by a synthesis (direct combination) reaction (see p. 157).

▶ 12 Quantitative chemistry

Objectives

When you have completed this chapter you should be able to:
- Understand **relative atomic mass**
- Understand **relative formula mass**
- Find the **percentages of elements** in a compound
- Understand the concept of the **mole**
- Find the **number of moles** in the mass of a substance
- Understand an **empirical formula**
- Work out **simple empirical formulae**
- Work out the **percentage composition** of a compound
- Find the **molecular formula** of a simple compound
- Calculate the **mass** of a product
- Calculate the **concentration** of solutions
- Understand **titrations** and the calculations associated with them
- Calculate the **volumes** of gases taking part in reactions.

The early chemists performed their experiments by trial and error. They did not calculate accurately how the quantities of chemicals reacted together. As chemists studied matter more closely they developed ways of making accurate calculations. Today these calculations are used in pharmacy, industry and in every laboratory where chemicals are used or studied.

The mass of an atom is far too small to use in calculations so the concept of relative atomic mass was developed. (This is introduced on p. 57.)

12.1 The relative atomic mass

The mass of the atom of one element is different to the masses of atoms of the other elements. The relative atomic mass was devised so that the masses of the atoms could be compared. The symbol for the relative atomic mass is Ar.

12.2 Relative formula masses

Relative atomic masses can be used to work out the relative formula mass or RFM of a compound. This is calculated by finding the Ar of each atom in the compound and adding them all together.

EXAMPLE

Copper sulphate has the formula Cu SO_4

From Table 4.4 (p. 57) it can be seen that the Ar of copper is 63.5, the Ar of sulphur is 32 and the Ar of oxygen is 16.

The RFM of copper sulphate is:

from the one atom of copper	63.5
from the one atom of sulphur	32.0
from the four atoms of oxygen	64.0
Total	159.5

QUESTION

1 What is the RFM of each of the following compounds (a) Ammonia NH_3, (b) Magnesium oxide MgO, (c) sodium chloride NaCl, (d) sodium hydrogen carbonate $NaHCO_3$, (e) sulphuric acid H_2SO_4. (Use Table 4.4, p. 57 to help you.)

12.3 The RFM and the percentage of an element in a compound

The RFM can be used to find the percentage by mass of an element in a compound.

EXAMPLE

The percentage of calcium in calcium carbonate is found by:

- Writing the formula $CaCO_3$
- Finding the Ar of each element Ca = 40, C = 12, O = 16
- Finding the RFM of calcium carbonate:

from one atom of calcium	40
from one atom of carbon	12
from three atoms of oxygen	48
Total	100

- Dividing the Ar of calcium by the RFM of calcium carbonate

$40/100 = 0.4$

- Multiplying by 100 to find the percentage $100 \times 0.4 = 40\%$

QUESTION

2 Find the percentage of (a) carbon in calcium carbonate, (b) oxygen in calcium carbonate, (c) nitrogen in ammonia (NH_3), (d) sodium in sodium chloride (NaCl), (e) sulphur in sulphuric acid (H_2SO_4).

12.4 The quantities of chemicals taking part in a reaction

(i) Building up the concept of the mole: I

While the Ar and RFM have their uses, chemists found that they needed an extra concept to accurately calculate the numbers of atoms taking part in a reaction. The concept that they use is the **mole**. The concept of the mole is built up this way.

The number of atoms in 12 g of carbon 12 is 6.02×10^{23}. This large number is called the **Avogadro number**. (It is named after Amedo Avogadro, 1776–1856, he was a chemist whose studies on gases helped to understand how quantities of chemicals reacted together.)

The same number of atoms present in a mass of another element can be found by considering the **relative atomic mass** of the element.

EXAMPLE

The relative atomic mass of magnesium is 24 compared to 12 of carbon 12. This means that 24 g of magnesium will contain as same number of atoms as 12 g of carbon 12.

The relative atomic mass of sulphur is 32 compared to 12 of carbon 12. This means that 32 g of sulphur will contain the same number of atoms as 12 g of carbon 12.

At this stage, the concept of the mole can be stated as the amount of an element that has the same number of atoms as 12 g of carbon 12.

(ii) The molar mass

The relative atomic mass of an element when expressed in grams holds 6.02×10^{23} atoms and is called the **molar mass**. It may also be described as 1 mole.

The unit of molar mass is g/mol.

EXAMPLE

The molar mass of magnesium is 24 g/mol and the molar mass of sulphur is 32 g/mol. The same information may be stated as:

24 g of magnesium is 1 mole of magnesium and 32 g of sulphur is one mole of sulphur.

(iii) Building up the concept of the mole: 2

The concept of the mole can be taken further to include compounds.

EXAMPLE

The Ar of chlorine is 35.5 so the molar mass of one mole of chlorine atoms is 35.5 g/mol. A molecule of chlorine has two atoms (Cl_2) so the molar mass of one mole of chlorine molecules is $2 \times 35.5 = 71$ g/mol.

As elements and compounds can generally be called substances and as atoms and molecules can generally be called particles a general statement of the concept of the mole can be built up as follows:

A mole is the amount of a substance which has the same number of particles as the number of atoms in 12 g of carbon 12.

(iv) Using the mole concept

(a) Finding the number of moles in a mass of an element

The Ar of the substance is discovered and expressed as a molar mass in g/mol. The molar mass is then divided into the mass of the substance present.

EXAMPLES

   1 How many moles of calcium are there in 80 g of calcium?

The Ar of calcium is 40.
Its molar mass is 40 g/mol

The number of moles of calcium = mass of calcium = $\dfrac{80}{40}$

molar mass of calcium = 2 mol

   2 How many moles of aluminium in 7 g of aluminium?

The Ar of aluminium is 27.
Its molar mass is 27 g/mol

The number of moles of aluminium is $\dfrac{7}{27} = 0.26$ mol

   3 How many moles of oxygen in 90 g of oxygen gas?

The Ar of oxygen is 16
The formula for a molecule of oxygen gas is O_2
The molar mass of the oxygen molecule is $16 \times 2 = 32$ g/mol

The number of moles of oxygen is $\dfrac{90}{32} = 2.81$ mol

QUESTIONS

   3 How many moles of calcium are there in (a) 160 g of calcium, (b) 6 g of calcium?
   4 How many moles of aluminium are there in 135 g of aluminium?
   5 How many moles of carbon are there in 400 g of carbon? (C = 12)
   6 How many moles of nitrogen are there in 168 g of nitrogen? (N = 14. A molecule of nitrogen has 2 atoms.)

(b) Calculating the amount of substance needed for particular molar amounts

The amount to weigh out is found by finding the molar mass of a substance and multiplying it by the number of moles required.

EXAMPLE

If three moles of sulphur are needed:

The molar mass of sulphur is 32 g mol
The amount needed is $32 \times 3 = 96$ g

QUESTION

7   What is the amount of substance needed for (a) 1.5 moles of carbon ($C = 12$), (b) 14 moles of aluminium ($Al = 27$), (c) 0.5 mole of calcium ($Ca = 40$)?

(c) Working out the empirical formulae

Using the mole concept and masses

The empirical formula of a substance shows the **ratio of atoms present**. For example the empirical formula of carbon dioxide is CO_2. It shows that there are two oxygen atoms for each carbon atom in the compound. The empirical formula of calcium oxide is CaO. It shows that the ratio of calcium to oxygen ions in the compound is one to one.

The empirical formula of a compound can be worked out by weighing a reactant and its product and applying the mole concept.

EXAMPLE

When 2.5 g of magnesium is heated it forms 4.2 g of magnesium oxide.

This means that 2.5 g of magnesium has combined with $4.2 - 2.5 = 1.7$ g of oxygen

The number of moles of magnesium and oxygen that have formed magnesium oxide is:

$$\text{Magnesium (Ar} = 24) = \frac{2.5}{24} = 0.1$$

$$\text{Oxygen (Ar} = 16) = \frac{1.7}{16} = 0.1$$

0.1 mole of magnesium combines with 0.1 mole of oxygen
so 1 mole of magnesium combines with 1 mole of oxygen
so the empirical formula of magnesium oxide is MgO.

QUESTION

8   What is the empirical formula of the compound formed when (a) 12 g of carbon combines with 32 g of oxygen, (b) 112 g of iron combines with 64 g of sulphur, (c) 40 g of calcium reacts with 71 g of chlorine?

Ar C = 12, O = 16, Fe = 56, S = 32, Ca = 40, Cl = 35.5

(*Note*: If the formula is A_2B_2 or A_3B_3 it is simplified to just AB. Similarly A_2B_4 is simplified to AB_2.)

Using the mole concept and percentages

The empirical formula can also be found from the percentage composition (by mass) of the elements in a compound. This is done by dividing the percentage of the mass of the substance present with its molar mass.

EXAMPLE

A compound contains 50% sulphur and 50% oxygen

Ar S = 32, O = 16

The moles of sulphur in the compound are $\dfrac{50}{32} = 1.6$ moles

The moles of oxygen in the compound are $\dfrac{50}{16} = 3.2$ moles

1.6 moles of sulphur combine with 3.2 moles of oxygen.

There is twice as much oxygen in the compound as sulphur so the empirical formula is SO_2.

QUESTION

9 What is the empirical formula of a compound that is (a) 40% sulphur, 60% oxygen, (b) 30% nitrogen, 70% oxygen (Ar N = 14) (c) 34.5% iron, 65.5% chlorine (Ar Fe 56, Cl (35.5))?

(d) Finding the molecular formula

The empirical formula only shows the ratio of atoms. For example, we have seen that in the instructions to Question 8 results such as A_2B_2 can be simplified to AB. Similarly results such as A_2B_4 could be simplified to give a ratio of AB_2.

In a molecular formula the number of each kind of atom in a molecule can be worked out.

This is done by:

(1) knowing the molar mass of the compound
(2) finding the actual mass of the elements in the compound
(3) dividing the mass of each one by its molar mass
(4) comparing the moles of elements present in the compound to discover the empirical formula
(5) using the empirical formula to calculate the molar mass of a substance with the empirical formula
(6) dividing the actual molar mass of the substance by the molar mass of the empirical formula substance to find the real numbers of atoms in the compound and not the ratio.

EXAMPLE

(1) The molar mass of a compound was found to be 84
(2) The actual masses of the elements in the compound were found to be 12 g of carbon and 2 g of hydrogen

(3) Moles of carbon present $\dfrac{12}{12} = 1$

Moles of hydrogen present $\dfrac{2}{1} = 2$

(4) 1 mole of carbon is combined with 2 moles of hydrogen
The empirical formula is CH_2

(5) As $C = 12$ and $H = 1$ a substance which had this empirical formula as its real formula would have a molar mass of

$$12 = 2 \times 1 = 14$$

(6) $\dfrac{\text{The actual molar mass}}{\text{the molar mass of the empirical formula substance}} = \dfrac{84}{14} = 6$

The real substance has 6 times the number of carbon and hydrogen atoms of the empirical formula substance. This means that the real formula of the substance is C_6H_{12}.

QUESTION

10   A substance has a molar mass of 56 and a sample is found to contain 24g of carbon and 4g of hydrogen. What is the real formula of the substance? ($C = 12, H = 1$.)

(e) Finding the mass of a product

When calcium carbonate is heated it decomposes to calcium oxide and carbon dioxide. The mass of the calcium oxide produced in the reaction can be found in two different ways.

Using the mole concept

(1) The symbol equation is constructed.
(2) The number of moles of reactant and product are identified.
(3) The RFM of the reactant is calculated.
(4) The number of moles of reactant taking part is calculated from the actual mass of the reactant taking part.
(5) The number of moles of the product is worked out by examining the relationship between the moles of reactant used and the moles of product produced.

EXAMPLE

20g of calcium carbonate is heated until decomposition is complete.

(1) $CaCO_3$ (s) $\rightarrow$ CaO (s) + CO_2 (g)
(2) 1 mole of calcium carbonate produces 1 mole of calcium oxide and 1 mole of carbon dioxide
(3) The RFM of $CaCO_3$ is

$$(Ca = 40, C = 12, O = 16 \times 3) = 100$$

(4) If one mole was being decomposed it would have a mass of 100g, but only 20g is being decomposed so it has a mass of 20/100 = 0.2 mole.

(5) Since 1 mole of calcium carbonate produces one mole of calcium oxide then 0.2 mole of calcium carbonate produces 0.2 mole of calcium oxide

The RFM of CaO is

$$(Ca = 40, O = 16) = 56$$

Therefore 0.2 mole of calcium oxide has a mass of

$$(0.2 \times 56) = 11.2g$$

QUESTION

11 How much calcium oxide is produced when 50g of calcium carbonate is completely decomposed?

Using the RFMs of the reactant and product

(1) The RFM of the reactant is found.

(2) The RFM of the product is found.

(3) The relationship between the molar masses of the reactant and product are examined.

(4) The amount of product is calculated by:

$$\frac{\text{The RFM of the product} \times \text{The actual mass of the reactant}}{\text{The RFM of the reactant}}$$

EXAMPLE

20g of calcium carbonate is heated until decomposition is complete

$$CaCO_3 \text{ (s)} \rightarrow CaO \text{ (s)} + CO_2 \text{ (g)}$$

(1) The RFM of $CaCO_3$ is

$$(Ca = 40, C = 12, O = 16 \times 3) = 100$$

(2) The RFM of CaO is

$$(Ca = 40, O = 16) = 56$$

(3) 100g of calcium carbonate produces 56g of calcium oxide

(4) The amount of calcium oxide produced by 20g of calcium carbonate is

$$\frac{56 \times 20}{100} = 11.2 \text{ g}$$

QUESTION

12 How much calcium oxide is produced when 80g of calcium carbonate is completely decomposed?

(v) Moles in solutions

A solution is composed of a **solute** and a **solvent**. The solute is the substance that dissolves and the solvent is the liquid in which the solute dissolves. A molar solution is a solution which has one mole of the solute dissolved in one litre of the solvent. The unit in which a molar solution is measured is in moles/litre, $moll^{-1}$ or $mol\,dm^3$. The molarity of a solution is indicated by a number with a capital M after it. For example, $1\,M$ is a molar solution, $0.5\,M$ is a half molar solution.

(a) Concentration

The concentration of the solute in the solution can be found using the following equation:

$$\text{The concentration(mol/l)} = \frac{\text{Amount of solute in moles}}{\text{Volume of solution in litres}}$$

EXAMPLE

If one mole of solute is dissolved in 1 litre of solution

$$\text{The concentration is } \frac{1\,\text{mole}}{1\,\text{litre}} = 1\,\text{mol/l}$$

(b) Changing the concentration

How concentration changes with volume

If 1 mole of solute is dissolved in 2 litres of solution

$$\text{The concentration is} \frac{1\,\text{mole}}{2\,\text{litres}} = 0.5\,\text{mol/l}$$

If 1 mole of solute is dissolved in 500 ml of solution

$$\text{The concentration is} \frac{1\,\text{mole}}{0.5\,\text{litres}} = 2\,\text{mol/l}$$

Finding the concentration of a solution

The concentration of a solution in mol/l is worked out from the mass of the solute dissolved and the volume of solution made up.

EXAMPLE

If 117g of sodium chloride is dissolved in 1 litre of water the number of moles of sodium chloride is found by taking the RFM of sodium chloride and dividing it into the mass dissolved

RFM of sodium chloride NaCl (Na = 23, Cl = 35.5) is 58.5

Mass dissolved is 117

$$\text{So number of moles of sodium chloride} = \frac{117}{58.5} = 2$$

The concentration is

$$\frac{\text{The amount of sodium chloride in moles}}{\text{The volume of solution in litres}} = \frac{2}{1} = 2\,\text{mol/l}$$

If the same amount of sodium chloride is dissolved in 2 litres of solution the concentration is

$$\frac{2}{2} = 1\,mol/l$$

13 What would be the concentration of sodium chloride if 117g of it was dissolved in 500 ml of solution?

14 What is the concentration of a one litre solution which has 240g of magnesium sulphate ($MgSO_4$) dissolved in it (Mg = 24, S = 32, O = 16)?

Finding the amount of solute in a solution

The equation linking concentration, solute and solution may be rearranged to find the amount of solute in a solution.

Amount of solute (mol) = Volume of solution (l) × Concentration (mol/l)

EXAMPLE

The number of moles of sodium hydroxide in a litre of solution of concentration 2 mol/l is

1 (l) × 2 (mol/l) = 2 moles

The number of moles of sodium hydroxide in 0.5 litres of solution of concentration 6 mol/litre is

0.5 (l) × 6 (mol/l) = 3 moles

QUESTION

15 What is the amount of solute of a substance which has a concentration of (a) 5 mols/l in 0.25 l of solution, (b) 3 mols/l in 9 l of solution, (c) 2.5 mols/l in 4 litres of solution?

Finding the mass of a solute to add to a solution

The equation:

Amount of solute (mol) = Volume of solution (l) × Concentration (mol/l)

can also be used to find the mass of a substance that must be dissolved to provide a volume of solution with a particular concentration.

EXAMPLE

If a litre of sodium hydroxide of concentration 2 mol/l is required then the amount of sodium hydroxide needed is

1 × 2 = 2 moles

The RFM of sodium hydroxide is

Na = 23, O = 16, H = 1, total = 40

so the mass of sodium hydroxide needed is

$$2 \times 40 = 80\,g$$

QUESTIONS

16 What amount of sodium hydroxide (NaOH) is needed (a) for 4 litres of solution at 3 mol/l concentration, (b) 6 litres of solution at 5 mol/l concentration? (Na = 23, O = 16, H = 1.)

17 What amount of potassium hydroxide (KOH) is needed for 250 ml of solution at a concentration of 5 mol/l? (K = 39, O = 16, H = 1.)

12.5 Titrations

Titrations are used in neutralisation reactions (see p. 155). They are also used to find the molarity of acids and alkalis.

For example, the following procedure can be used to find the molarity of a sample of sulphuric acid:

(1) An alkali of known molarity such as 1.0 M sodium hydroxide solution is selected.
(2) 25 ml of the 1.0 M sodium hydroxide solution is measured into a conical flask.
(3) The acid of unknown molarity is carefully poured into a burette using a filter funnel until the zero mark at the top of the burette is reached.
(4) A few drops of an indicator such as phenolphthalein are added to the sodium hydroxide solution. The indicator colours the solution.
(5) The conical flask is placed under the burette and the acid is added 0.5 ml at a time. Between each addition the mixture is swirled and the colour of the indicator is checked. As the indicator begins to change smaller volumes of acid are added until the indicator changes colour (or in the case of phenolphthalein the colour disappears). This change marks the end point of the titration and the volume of acid is read from the scale on the burette. The procedure should be repeated to ensure consistent results.
(6) For neutralisation to take place the number of acid 'particles' must be the same as the number of alkali 'particles' (see pp. 151 and 153 to consider the ions involved). This balance in the number of ' particles' is expressed by the equation:

$$\frac{M_1 V_1}{Macid} = \frac{M_2 V_2}{Malkali}$$

where M_1 = the molarity of the acid
    V_1 = the volume of the acid
    Macid = number of moles of acid taking part in the reaction
    M_2 = the molarity of the alkali
    V_2 = the volume of the alkali
    Malkali = number of moles of alkali taking part in the reaction.

Macid and Malkali are found from the symbol equation for the reaction. From the information provided so far, we can fill in parts of the equation:

$$\frac{M_1V_1}{Macid} = \frac{1 \times 25}{Malkali}$$

Macid and Malkali can be discovered by constructing the symbol equation and counting the number of moles of the acid and alkali taking part in the reaction:

$$2NaOH \text{ (aq)} + H_2SO_4 \text{ (aq)} \rightarrow Na_2SO_4 \text{ (aq)} + H_2O \text{ (l)}$$
$$\quad 2 \text{ moles} \qquad 1 \text{ mole}$$

This shows that Macid is 1 and Malkali is 2.

The equation can be modified to:

$$\frac{M_1V_1}{1} = \frac{1 \times 25}{2}$$

The volume of acid used in the neutralisation reaction is 20 ml, so the equation can be further modified to

$$\frac{M_1 \times 20}{1} = \frac{1 \times 25}{2}$$

Rearranging the equation to discover M_1:

$$M_1 = \frac{1 \times 25 \times 1}{2 \times 20} = \frac{25}{40} = 0.625 \text{ M}$$

The amount of suphuric acid can be expressed in grams per litre by multiplying the molarity of the acid with its RFM:

$$0.625 \times 98 = 61.25 \text{ g/litre of solution}$$

QUESTIONS

18 What is the molarity of 25 ml sodium hydroxide solution which is neutralised by 20 ml of 3.0 M sulphuric acid solution?

19 What is the molarity of 30 ml of hydrochloric acid which has been neutralised by 25 ml of 2 M sodium hydroxide solution? (HCl (aq) + NaOH (aq) $\rightarrow$ NaCl (aq) + H_2O (l).)

12.6 Moles and gases

Avogadro (see p. 161) established a law about the number of molecules in equal volumes of gas that are kept in the same conditions. The law states that **equal volumes of gas kept in the same conditions contain the same number of molecules**.

From this law it can be seen that one mole of one gas at room temperature and pressure will occupy the same volume as a sample of another gas in the same conditions. The volume occupied by a mole of gas at room temperature and pressure (abbreviated to rtp) is 24 dm^3. This volume is called the **molar gas volume**. The volume occupied by 0.5 moles of a gas at rtp is $0.5 \times 24 = 12 \text{ dm}^3$ and the volume occupied by 6 moles of a gas at rtp is $6 \times 24 = 144 \text{ dm}^3$.

QUESTION

    20   What is the volume occupied by (a) 0.2 M, (b) 3.0 M ammonia gas at rtp?

(i) Finding the volumes of gases involved in a chemical reaction where all the substances are gases

- Write down the word equation
  For example:

    Hydrogen + chlorine $\rightarrow$ hydrogen chloride

- Write down the symbol equation to find the number of moles of gas in the equation:

    H_2 + Cl_2 $\rightarrow$ 2HCl
  one mole   one mole   two moles

For each mole substitute the volume 24 dm³:

    24 dm³ hydrogen = 24 dm³ chlorine $\rightarrow$ 48 dm³ hydrogen chloride

(ii) Finding the volume of a gas evolved from a solid

- Write down the word equation
  For example:

    calcium carbonate $\rightarrow$ calcium oxide + carbon dioxide

- Write down the symbol equation:

    $CaCO_3$ $\rightarrow$ CaO   + CO_2
  one mole of   one mole of   one mole of
  calcium carbonate   calcium oxide   carbon dioxide

    40 + 12 + 48 = 100 g of calcium carbonate produces 24 dm³ of carbon dioxide

If the mass of calcium carbonate was only 10 g the volume of gas produced is calculated by

$$\text{Volume} = \frac{\text{Mass of solid reactant}}{\text{Mole of solid reactant}} \times 24\,\text{dm}^3$$

so with 10 g of calcium carbonate

$$\text{The volume of carbon dioxide} = \frac{10 \times 24\,\text{dm}^3}{100} \times 2.4\,\text{dm}^3$$

QUESTIONS

    21   What volume of carbon dioxide would be produced if (a) 50 g, (b) 200 g of calcium carbonate was decomposed? (Ar Ca = 40, O = 16, C = 12.)
    22   What volume of carbon dioxide is produced when 24 g of carbon is burnt in oxygen? (Ar C = 12, O = 16.)

12.7 Summary

- The relative atomic mass is used to compare the atoms of different elements (see p. 159).
- The relative formula mass is derived from the relative atomic masses of the different elements in a compound (see p. 159).
- The relative formula mass can be used to find the percentages of elements in a compound (see p. 160).
- The mole is an amount of an element or compound that has the same number of atoms or molecules as 12 g of carbon 12 (see p. 162).
- The molar mass of a substance is its relative atomic mass or relative formula mass weighed out in grams. (see p. 161).
- The mole concept allows many calculations to be made (see p. 162).
- The number of moles in a mass of a substance can be calculated (see p. 162).
- The mole concept can be used to calculate the amount of a substance needed for a particular reaction (see p. 162).
- The empivical formula of a compound can be worked out (see p. 163).
- The molecular formula of a compound can be worked out (see p. 164).
- The mass of a product in a reaction can be calculated (see p. 165).
- The concentrations of solutions can be calculated (see p. 166).
- Titrations can be used to find the concentrations of acids and alkalis (see p. 169).
- The volumes of gases which take part in a chemical reaction can be calculated (see p. 170).

■ M̌ **1 3** Electrochemistry

Objectives

When you have completed this chapter you should be able to:
- Understand how to set up a **circuit for electrolysis**
- Identify two types of **electrode**
- Explain the **reactions** at the electrodes
- Understand how water affects the **electrolysis of aqueous solutions**
- Calculate the **amounts of elements** deposited on the electrodes
- Calculate the **volume of gas** produced at an electrode
- Use the **Faraday** in calculations
- Explain the **electroplating process**
- Describe how an aluminium object is **anodised**.

13.1 Conductors and non-conductors

Materials can be divided into two groups according to whether or not they are able to allow a current of electricity to pass through them. Those materials which allow a current of electricity to pass through them are called **conductors** (see p. 107). Those materials which do not allow a current of electricity to pass through them are called **non-conductors** (see pp. 65 and 70).

Metals are conductors of electricity. The atoms of metals are surrounded by a 'sea of electrons' (see p. 70). These electrons carry the current through the metal and no chemical change occurs as the current passes through.

There are some conductors which allow a current of electricity to pass through them but as they do so they undergo a chemical decomposition. These conductors are called **electrolytes**. There are two kinds of electrolytes – electrolytes made of molten ionic compounds and electrolytes made from ionic substances dissolved in water. Both kinds of electrolytes are decomposed by setting up a circuit as described on p. 174. The process in which electricity is used to decompose a substance is called **electrolysis**. This process has several useful applications.

13.2 The circuit for electrolysis

The following components are needed for the circuit – a direct current source such as a battery, rods called electrodes to dip into the electrolyte, wires to

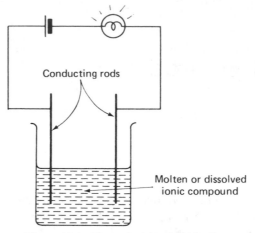

Figure 13.1 Electrolysis using a beaker

connect the current source to the electrodes, a switch and a container for the electrolyte. The container may range in structure from a crucible to a Hofmann voltameter (see Figure 13.3, p. 179). In Figure 13.1, it is shown in its simplest form – a beaker.

The electrode connected to the positive terminal of the battery is called an **anode**.

The electrode connected to the negative terminal of the battery is called the **cathode**.

The electrodes are made of materials which do not decompose when electricity passes through them (but see electroplating, p. 183). They may be made of the non-metal graphite or metals such as platinum, stainless steel or titanium. The electrodes for a particular electrolysis reaction are chosen so that they do not react with the electrolyte or the products. For example, in the electrolysis of lead bromide (see p. 175) the anode is made of graphite and the cathode is made of steel.

(i) When the current is switched on

The electrolyte contains positively charged ions and negatively charged ions which are free to move when the current is switched on.

(ii) At the anode

When the current is switched on the anode loses electrons to the battery and becomes positively charged. The negatively charged ions called **anions** are attracted to the anode and when they reach it they give up the electrons which gave them their charge. When the ions lose their electrons they become neutral atoms or molecules. When this occurs the anions are said to have been discharged.

(iii) At the cathode

When the current is switched on the cathode gains electrons from the battery and becomes negatively charged. The positively charged ions called **cations** are attracted to the cathode and when they reach it they take up electrons and become neutral atoms or molecules. When this occurs the cations are said to have been discharged.

QUESTIONS

1 How are electrolytes different from other conductors?
2 Distinguish between anions and cations.
3 How is the flow of electrons round the circuit maintained by the electrolyte?

13.3 Molten electrolytes

(i) Lead bromide

Lead bromide is a compound that is often used to demonstrate electrolysis in a molten electrolyte. The lead bromide is heated in a crucible above its melting point of 373°C. When the current of electricity is switched on the bromide ions move to the anode and the lead ions move to the cathode. The changes which take place at the electrodes can be shown as half equations (see also p. 81).

(a) At the anode

Two bromide ions – two electrons → one bromine molecule
$2Br^- (l) - 2e^- \rightarrow Br_2 (g)$

The bromine is released as a gas as its boiling point of 59°C is well below the temperature of the molten electrolyte.

(b) At the cathode

one lead ion + two electrons → one lead atom

$Pb2^+ (l) + 2e^- \rightarrow Pb (l)$

The lead is released as a liquid as its melting point of 323°C is below the temperature of the molten electrolyte.

When the two half equations are put together the electrons given up by the bromide ions are balanced by those taken up by the lead ion and they are removed from the combined equation:

$PbBr_2 (l) \rightarrow Pb (l) + Br_2 (g)$

QUESTION

4 Look at the descriptions of oxidation (p. 113) and reduction (p. 114) then identify where they occur in electrolysis.

In the lead bromide demonstration the electrolysis may have taken place in a fume cupboard to remove the harmful fumes. Alternatively it may have taken place using a flask with a side arm through which the gas passed for safe collection.

(ii) Sodium chloride

Sodium is obtained by the electrolysis of molten sodium chloride in a Downs cell (Figure 13.2).

The cell is modified for the efficient collection of the products of decomposition. The chlorine gas released at the central graphite anode collects in a funnel and is drawn away for use in industry. The steel cathode forms a circular ring around the funnel in which the molten sodium metal collects.

The half equations for the reactions which take place at the electrodes are:

(a) At the anode

two chloride ions – two electrons → one chlorine molecule

$2Cl^- (l) - 2e^- \rightarrow Cl_2 (g)$

(b) At the cathode

sodium ion + one electron → one sodium atom

$Na^+ (l) + 1e^- \rightarrow Na (l)$

To balance the equations when they are combined the half equation at the cathode must be doubled to:

$2Na^+ (l) + 2e^- \rightarrow 2Na (l)$

so the combined equation is:

$2NaCl (l) \rightarrow 2Na (l) + Cl_2 (g)$

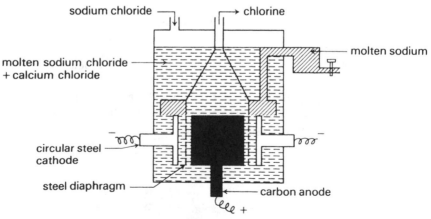

Figure 13.2 A Downs cell

13.4 Aqueous solutions

The ions that form in the water can be arranged in the order in which they will be released from the electrolyte. The ions of the least reactive metals are released before the more reactive metals, as Table 13.1 shows.

Table 13.2 shows the order in which anions are most easily released or discharged. The ions of carbonates, nitrates and sulphates are not discharged.

Note: When hydroxide ions are discharged they very quickly form water and oxygen at the anode:

$$OH^- - e^- \rightarrow OH \text{ (aq)}$$
$$4OH \text{ (aq)} \rightarrow 2H_2O \text{ (l)} + O_2 \text{ (g)}$$

QUESTIONS

5 How does Table 13.1 compare with Table 8.1?
6 How does Table 13.2 compare with the table you constructed in answer to Question 19 in Chapter 9 (p. 134)?

silver	Ag^+
copper	Cu^{2+}
lead	Pb^{2+}
tin	Sn^{2+}
zinc	Zn^{2+}
hydrogen	H^+
aluminium	Al^{3+}
magnesium	Mg^{2+}
calcium	Ca^{2+}
sodium	Na^+
potassium	K^+

Table 13.1   Order of release or discharge of cations

iodide	I^-
bromide	Br^-
chloride	Cl^-
oxide/hydroxide	O^{2-}/OH^-

Note: When hydroxide ions are discharged they very quickly form water and oxygen at the anode:

$$OH^- - e^- \rightarrow OH \text{ (aq)}$$
$$4 OH \text{ (aq)} \rightarrow 2H_2O \text{ (l)} + O_2 \text{ (g)}$$

Table 13.2   Order of release or discharge of anions

13.5 Water – an electrolyte

When an ionic compound dissolves in water it forms ions just as it would do if it were molten. However, water is also an electrolyte too although it is only a very weak one. The ions that it produces are shown in the equation:

$$H_2O \text{ (l)} \rightleftharpoons H^+ \text{ (aq)} + OH^- \text{ (aq)}$$

There are only a small number of ions present compared to the huge number of water molecules but the double arrow shows that a balance is made between the ions and the molecules. This means that if the ions are discharged more water molecules break up to form ions to replace them. The effect of this is to make water seem as if it is made up of a huge number of ions.

(i) The electrolysis of water

Water can be decomposed by electricity into hydrogen and oxygen gases. A little sulphuric acid is added to the water to speed up the reaction and the water is set up in the apparatus called a Hofmann voltameter, as shown in Figure 13.3.

QUESTION

7 Look at the volumes of the gases in the Hofmann voltameter. How do you think the result supports the formula of water – H_2O?

By looking at Tables 13.1 and 13.2 you can see that hydrogen ions are half way down the list of cations and hydroxide ions are at the bottom of the list of anions. This means that if a cation below hydrogen is present in the water hydrogen from water molecules will be discharged instead of the cation. It also means that if an anion above the hydroxide ion is present in the water that anion will be released instead of the hydroxide ion. If carbonate, nitrate or sulphate ions are present the hydroxide ion will be discharged because these ions are not in the list and are not discharged at the anode.

(ii) Electrolysis of sodium chloride solution

A solution of sodium chloride contains four ions – Na^+, H^+, Cl^- and OH^-. By looking at Tables 13.1 and 13.2 you can tell which ions will be discharged at the electrodes and which will remain in solution.
   At the anode, chlorine gas is released:

$$2Cl^- \text{ (aq)} - 2e^- \rightarrow Cl_2 \text{ (g)}$$

At the cathode hydrogen gas will be released:

$$2H^+ \text{ (aq)} + 2e^- \rightarrow H_2 \text{ (g)}$$

The word equation for the whole reaction is:

sodium chloride + water → sodium ions + hydroxide ions + hydrogen + chlorine

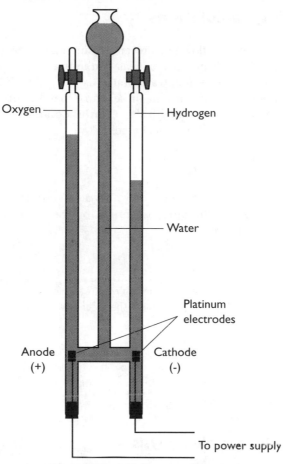

Figure 13.3   A Hofmann voltameter

The sodium and hydroxide ions form a solution of sodium hydroxide.

Figure 7.2 (p. 94) shows a membrane cell which is used in industry for the electrolysis of brine to produce hydrogen, chlorine and sodium hydroxide solution.

QUESTIONS

8 Produce a symbol equation for the word equation describing the electrolysis of sodium chloride solution

9 Produce the half equations occurring at the electrodes for the electrolysis of (a) zinc bromide solution ($ZnBr_2$ (aq)), (b) copper sulphate solution ($CuSO_4$ (aq)).

10 Where are hydrogen and chlorine gases produced in the electrolysis of hydrochloric acid?

13.6 Making calculations

Huge numbers of electrons flow in a current through an electrolyte when electrolysis is taking place. This number is measured in units called **amps**. The total number of electrons flowing through a circuit in a certain time is found by multiplying the current (in amps) by the time (in seconds) that the current flowed. This unit of amp second is called a coulomb after **Charles Coulomb** (1736–1806), who devised a way of measuring electrical charge.

(i) Calculating coulombs

The coulombs of electricity which flow through an electrolyte in a certain time are found by multiplying the amps (recorded by the ammeter) by the time (recorded by the stop clock).

EXAMPLE

When a 2 amp current passes through an electrolyte for three minutes $2 \times 3 \times 60$ = 360 coulombs of charge pass through the electrolyte.

QUESTION

11 How many coulombs of electricity flow through an electrolyte when a current of (a) 1 amp flows for 30 seconds, (b) 1 amp flows for 1 minute, (c) 2 amps flows for 15 minutes?

(ii) Relationships in electrolysis

Simple investigations in which different amounts of electricity are used in electrolysis reactions will show that the amount of element produced at an electrode depends on the amount of electricity that flows through the circuit. This can be stated as: **the amount of element** produced at an electrode is directly proportional to the amount of electricity which has flowed through the electrolyte.

The amounts of elements produced at the electrodes can be calculated by examining the half equation (see p. 81). Which describes the reaction taking place there.

EXAMPLE

In the electrolysis of copper sulphate the half equation describing the reaction at the cathode is

one copper ion + 2 electrons → one copper atom
Cu^{2+} (aq) + $2e^-$ → Cu (s)

Applying the concept of the mole (see p. 161), it can be seen that

1 mole of copper ions + 2 moles of electrons
    → one mole of copper atoms

The charge on one mole of electrons the faraday is 96.500 coulombs. This information can be used to modify the equation in the following way:

1 mole of copper ions + 2 × 96 500 coulombs → 1 mole of copper atoms

The equation can then be rearranged to find how much copper one coulomb of electricity can discharge at the cathode.

$$\text{The amount} = \frac{1 \text{ mole of copper atoms}}{2 \times 96\,500}$$

As 1 mole of copper atoms in grams is equivalent to the Ar of copper (see p. 57), the equation can be further changed to

$$\text{The amount} = \frac{Ar \text{ of copper}}{2 \times 96\,500}$$

The equation can be further generalised by

$$\text{The amount of an element} = \frac{Ar \text{ of element}}{\text{Charge on the ion} \times 96\,500}$$

The mass of an element discharged when a current of any number of coulombs of electricity is passed through the electrolyte is found by

$$\text{Mass of element} = \frac{Ar \text{ of element} \times Number \text{ of coulombs}}{\text{Charge on the ion} \times 96\,500}$$

QUESTION

12   Use the general equation to calculate the mass of (a) copper (Ar = 63.5) produced at an electrode when a current of 2 amps passes through the electrolyte for 4 minutes, (b) silver (Ar = 108) produced at an electrode when a current of 1 amp passes through the electrolyte for 30 seconds.

The general equation for masses of elements produced at electrodes can be used for both cations and anions. When a solution of copper chloride is electrolysed the solid metal copper is deposited at the cathode and the gaseous chlorine is released at the anode.

QUESTION

13   What mass of chlorine (Ar = 35.5) is produced at an anode when a current of 3 amps flows through the copper chloride solution for one minute?

Chemists are usually more interested in the **volume** of a gas that is produced rather than its mass as this helps them to calculate the space needed to collect it at rtp or to predict equipment needed to store it under pressure or at other temperatures.

The way the mass of a gas is converted to a volume is shown in the following equation:

$$\text{The volume of gas} = \frac{\textbf{Mass of gas produced}}{\textbf{Mass of one mole of gas}} \times 24000 \text{ cm}^3$$

14   In an electrolysis experiment 0.1 g of chlorine is produced. What volume does this gas occupy at rtp? (Mass of one mole of gas = 35.5 × 2 = 71 g).

13.7 The Faraday

The charge on one mole of electrons is 96 500 coulombs. This unit is called the Faraday after Michael Faraday who performed many experiments on electricity. It is also used in electrochemical calculations. It is the **quantity of electricity which will release one mole of a substance with one electrical charge on its ions or half a mole of a substance with two electrical charges on its ions**.

EXAMPLE

Silver ions have one positive charge and an Ar of 108. One Faraday will liberate 108 g of silver at a cathode.

Copper ions have two positive charges and an Ar of 63.5. One Faraday will liberate 31.75 g of copper at a cathode or two Faradays will liberate 1 mole – 63.5 g.

QUESTION

15   Aluminium ions have three positive charges and an Ar of 27. How many Faradays are required to release a mole of aluminium at a cathode?

When the liberation of gases is considered it is important to remember that many gases have diatomic molecules. For example, chloride ions Cl^- form chlorine gas Cl_2. One mole of chloride ions are liberated by one Faraday but one mole of chlorine gas needs two Faradays to form.

QUESTION

16   Oxygen ions O^{2-} form molecules with a formula O_2. How many Faradays are needed to liberate one mole of gas (a) 1, (b) 2, (c) 4, (d) 6?

Usually only a fraction of a mole is liberated at an electrode but this can be used to calculate the quantity of electricity in Faradays that has been consumed.

EXAMPLE

20 g of silver was produced at a cathode.

The amount of electricity used was $\frac{20}{108}$ = 0.19 Faradays

The equation can be rearranged to find the amount of a substance produced by a quantity of electricity.

QUESTIONS

17   How many Faradays produce 54 g of silver?
18   How many grams of silver are produced by 0.75 Faradays?

13.8 Electroplating

This is a process in which electrolysis is used to give an object a metal coating from the electrolyte. The electrodes which are used in this process are an anode made of the metal that is in the electrolyte and a cathode made from the object to be coated, as shown in Figure 13.4.

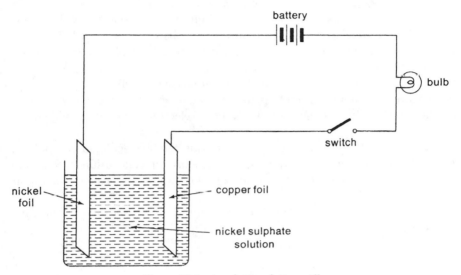

Figure 13.4 An electroplating cell

Electroplating has a variety of uses. It may be used to cover the connections of a micro-processor with gold to improve electrical conductance, to cover cutlery with silver to improve its appearance, to cover steel bath taps with chrome to prevent rusting.

When an object is being gold plated, for example, the following half equation shows the reaction taking place at the cathode:

gold ion + electron → gold atom
$Au^+ (aq) + 1e^- → Au (s)$

Electroplating is also used for the purification of copper (see p. 226).

13.9 Anodising

This electrolysis process is used to increase the layer of aluminium oxide on the surface of aluminium objects. The surface of aluminium reacts with oxygen in the air to produce a layer of aluminium oxide which prevents the metal from corrosion. Anodising gives the metal further protection and keeps its surface shiny.

In anodising, the aluminium object is used as the anode and the electrolyte is sulphuric acid. When electricity passes through the electrolyte oxygen forms at the anode and reacts with the metal to form a thicker protective coat.

19 Compare electroplating and anodising.

13.10 Summary

- A simple circuit is needed for electrolysis (see p. 173).
- There are two types of electrolyte – molten ionic compounds and aqueous solutions in which the ionic compounds have dissolved (see pp. 175–7).
- The reactions that take place at the electrodes can be explained by half equations (see p. 175).
- Water produces hydrogen and hydroxide ions (see p. 178).
- The ions in an aqueous solution are released in an order (see p. 177).
- The amount of an element produced at an electrode can be calculated (see p. 180).
- The volume of a gas produced at an electrode can be calculated (see p. 180).
- The Faraday is a unit of charge which can be used in calculations (see p. 182).
- Some metals are coated with a second metal by a process called electroplating (see p. 183).
- The layer of aluminium oxide on aluminium is increased by the anodising process (see p. 183).

▣ Ⓜ **14** Rates of reaction

Objectives

When you have completed this chapter you should be able to:
- Explain how particle size, concentration, temperature, catalysts and light affect the **rate of a reaction**
- Describe the properties of **a catalyst**
- Understand that **enzymes** are biological catalysts
- Explain reactions by means of the **collision theory**
- Describe non-reversible and reversible **physical changes**
- Describe non-reversible and reversible **chemical changes**
- Understand **equilibrium**
- Distinguish between **exothermic** and **endothermic** reactions
- Describe **activation energy**
- Describe how **bond energies** can be used in the study of a reaction
- Describe how the **amount of energy in a fuel** can be measured.

The rate of a reaction is the speed at which a reaction takes place. Some reactions take place very quickly. The reaction taking place after a firework is set off is an example of a very fast reaction. A few reactions take a long time. Concrete, for example, takes two days to set and fruit may take weeks to ripen. The rate of a reaction is affected by several factors. They are particle size, concentration, temperature, and the effect of a catalyst. For a few reactions light is an important factor.

14.1 The effect of particle size

The effect of this factor may be investigated with calcium carbonate and hydrochloric acid:

> calcium carbonate + hydrochloric acid → calcium chloride + water
> + carbon dioxide

> $CaCO_3 \text{ (s)} + 2HCl \text{ (aq)} \rightarrow CaCl_2 \text{ (aq)} + H_2O \text{ (l)} + CO_2 \text{ (g)}$

A mass of large marble chips is placed in a conical flask and a certain volume of 2M acid (more than is needed for the reaction to be completed) is added.

The rate of reaction may be recorded by measuring the change in mass of the reactants using the apparatus in Figure 14.1 or by measuring the volume of gas produced, as shown in Figure 14.2.

The mass or volume is measured every 30 seconds until the reaction is complete. The results of the investigation can be plotted on a graph, as shown in Figure 14.3.

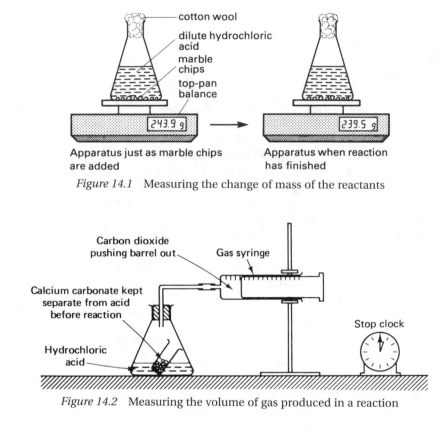

Figure 14.1 Measuring the change of mass of the reactants

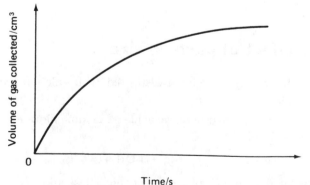

Figure 14.2 Measuring the volume of gas produced in a reaction

Figure 14.3 Measuring gas volume to show the rate of the reaction

The graph shows how the rate of reaction changes over time. The slope indicates the rate of reaction. For example, a steep slope at the beginning of the reaction shows that the rate of reaction is greater than later when the slope is less steep.

The investigation shown in Figure 14.1 is continued by repeating the experiment with the same mass of smaller marble chips and the same volume of 2M hydrochloric acid. When this has been done a second graph can be plotted on the first one (see Figure 14.4).

QUESTIONS

1 Why was more acid added than was needed? (*Hint:* Think about what would have happened if too little acid was added.)
2 How did the use of smaller particles affect (a) the rate of reaction, (b) the quantity of product? How does Figure 14.4 support your answers?
3 Make a sketch of Figure 14.4 and add the result you may expect if powdered limestone was used as the third part of the investigation. Explain your answer.
4 The investigation with marble chips and hydrochloric acid was repeated with magnesium ribbon, magnesium powder and hydrochloric acid. (a) Write a symbol equation for the reaction. (b) What results would you predict? (c) Explain your answer. (Magnesium = Mg.)

14.2 Particle size and surface area

The surface area of a particle depends upon its size. A 2 cm cube has six surfaces each with an area of $2 \times 2 = 4\,cm^2$. The total surface area of the cube is $6 \times 4 = 24\,cm^2$.

If the cube is cut into eight 1 cm cubes (see Figure 14.5), each of the small cubes has a side with a surface of $1\,cm^2$. This gives a total surface area for each cube of $6 \times 1 = 6\,cm^2$. The total surface area of the solid now is $6 \times 8 = 48\,cm^2$.

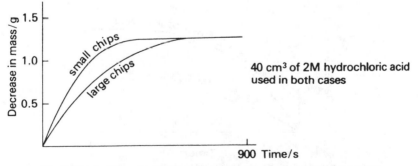

Figure 14.4 The effect of marble chip size on the rate of reaction

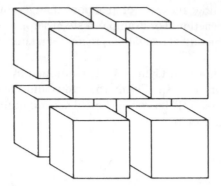

Figure 14.5 Cube cut into smaller cubes

QUESTIONS

    5 How does cutting a 2 cm cube into eight 1 cm cubes affect the area of the solid?

    6 How does the surface area of the solid change if each 1 cm cube is cut into eight ½ cm cubes?

14.3 The effect of concentration

The concentration of a solution is described on p. 167.

(i) Using marble chips and hydrochloric acid

The reactants can be set up in the apparatus shown in Figures 14.1 or 14.2 (p. 186).

    In the first part of the investigation a volume of $50 \, cm^3$ 2M acid is used as a reactant.

    In the second part of the investigation a volume of $25 \, cm^3$ 2M acid and $25 \, cm^3$ distilled water is used as a reactant (Figure 14.6).

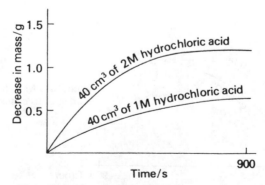

Figure 14.6 The effect of concentration change on the rate of reaction

7 How does diluting the acid affect the rate of the reaction?

(ii) Using sodium thiosulphate and hydrochloric acid

Sodium thiosulphate is soluble in water but when hydrochloric acid is added to the solution sodium chloride and water are formed, sulphur is precipitated and sulphur dioxide gas is produced.

The symbol equation for this reaction is:

$$Na_2SO_3 \text{ (aq)} + 2HCl \text{ (aq)} \rightarrow 2NaCl \text{ (aq)} + H_2O \text{ (l)} + S \text{ (s)} + SO_2 \text{ (g)}$$

This reaction is particularly useful because the precipitation of sulphur can be used as a measure of the rate of reaction.

The reactants are set up in a flask which is positioned over a paper with a cross on it (see Figure 14.7).

The reactants are viewed from above the neck of the flask. The speed of the reaction is measured by the time from the mixing the reactants to the time when the cross is no longer visible due to the precipitation of the sulphur.

The effect of concentration of the reactants may be investigated by using the same concentration of acid and varying the concentration of the sodium thio-sulphate solution or by using the same concentration of sodium thiosulphate solution and varying the concentration of the acid.

Figure 14.8 shows how the time for the reaction to be completed varys with the concentration of the sodium thiosulphate solution.

QUESTIONS

8 How are the axes of the graph in Figure 14.8 different from those in Figure 14.4?

9 How does the concentration of one of the reactants affect the rate of reaction?

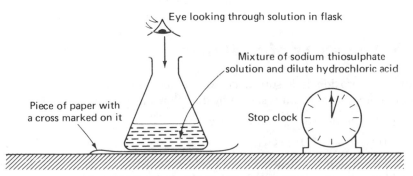

Eye looking through solution in flask

Mixture of sodium thiosulphate solution and dilute hydrochloric acid

Piece of paper with a cross marked on it

Stop clock

Figure 14.7 Conical flask with cross and paper

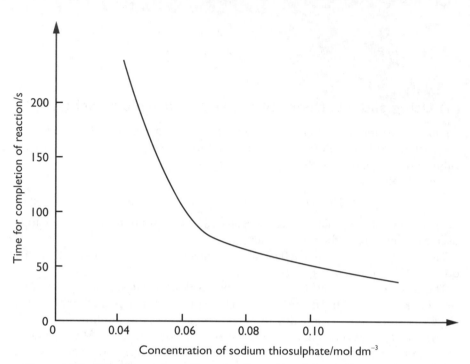

Figure 14.8 How the concentration of sodium thiosulphate affects reaction time

14.4 The effect of temperature

Sodium thiosulphate solution and hydrochloric acid can be used to investigate the effect of temperature. The same concentration of reactants are used throughout the investigation but the temperature of the reactants is varied.

Figure 14.9 shows how the rate of reaction changes by varying the temperature of the reactants.

QUESTION

    10 How does the temperature affect the rate of reaction?

14.5 The effect of a catalyst

Hydrogen peroxide is a substance which decomposes very slowly at room temperature. The reaction is shown by the following equations:

    hydrogen peroxide $\rightarrow$ water + oxygen

    $2H_2O_2 \, (l) \rightarrow 2H_2O \, (l) + O_2 \, (g)$

If a small amount manganese (IV) oxide is added to a sample of hydrogen peroxide the reaction is so fast that the mixture fizzes as the oxygen escapes.

The reaction can be investigated using the apparatus shown in Figure 14.2 (p. 186). As hydrogen peroxide decomposes so slowly on its own it would take a very long time to fill the syringe so the reaction can be investigated by using different amounts of the catalyst.

Figure 14.10 shows how the mass of a catalyst can affect the rate of reaction.

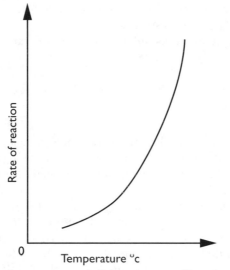

Figure 14.9 How the temperature of the reactants affects the rate of reaction

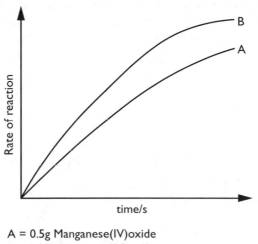

A = 0.5g Manganese(IV)oxide
B = 1.0g Manganese(IV)oxide

Figure 14.10 How the mass of catalyst affects the rate of reaction

14.6 Properties of catalysts

- Catalysts speed up reactions. (Chemicals which slow down reactions are called inhibitors.)
- At the end of a reaction a catalyst is not chemically changed and can be used again. A catalyst may physically change in a reaction. For example, manganese (IV) oxide forms smaller particles during the reaction with hydrogen peroxide.
- Only small amount of catalyst is needed to produce a large chemical reaction.
- Catalysts can work over a wide range of temperatures (see p. 240) and with gaseous reactants under pressure (see p. 238).
- A catalyst usually only works on one reaction. It is said to be **specific** to that reaction.

14.7 Enzymes

Enzymes are 'biological catalysts' They are made by living things and catalyse chemical reactions inside cells (and outside cells in the digestive system). These reactions keep the organism alive. Enzymes are made from protein molecules. Proteins are destroyed by heat so the temperature range over which an enzyme can work is much more limited than that of a 'chemical catalyst'. Figure 14.11 shows how the activity of an enzyme from the human body varies with temperature. Its activity is highest at body temperature. This temperature is called the optimum temperature for the enzyme.

The activity of an enzyme is also affected by the pH of its surroundings, as Figure 14.12 shows.

(i) Food and drink

An enzyme in the mouth only works in the alkalinity of the saliva there. When it is swallowed it becomes inactive in the acidity of the stomach where another enzyme continues the digestion process.

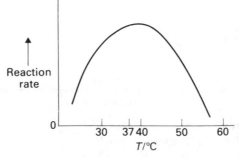

Figure 14.11 How the rate of an enzyme catalysed reaction varies with temperature

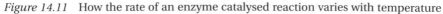

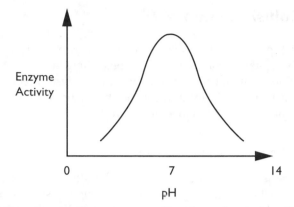

Figure 14.12 How the rate of an enzyme catalysed reaction varies with pH

Enzymes have been used for thousands of years to make food and drinks. Yeast produces an enzyme which breaks down sugar in bread-making and in the process of fermentation for the production of alcoholic drinks.

The equation of the fermentation process is:

glucose → ethanol + carbon dioxide

$C_6H_{12}O_6$ (aq) → $2C_2H_5OH$ (aq) + $2CO_2$ (g)

Enzymes are used in the production of cheese and yogurt.

(ii) Washing powders

Enzymes are also used in some washing powders (called biological washing powders), where they can remove stains made by food, sweat and blood at a lower temperature than non-biological powders. Some people are allergic to the enzymes in the biological powders and develop skin rashes. These can be avoided by using non-biological powders.

14.8 Light

The process of **photosynthesis** depends on light. In this reaction, carbon dioxide from the air and water from the soil react in the leaves and other green parts of a plant to produce carbohydrates and oxygen (see p. 112).

When you take a photograph light produces a chemical reaction in the film in your camera. In the film are crystals of silver bromide. When light strikes the crystals an electron from a bromide ion in the crystal is moved to a silver ion in the crystal and an atom of silver is produced. The small volume of bromine gas produced in this reaction is trapped in the material from which the film is made.

The image on the film is made by different amounts of light which produce areas on the film with different amounts of silver metal. These are used in the development of the film to produce the picture in the photograph. Silver is deposited in the making of both black and white and colour photographs.

14.9 Collision theory

In Chapter 1, p. 11 matter was described as being made from particles. The particle theory of matter can be used to explain how factors such as surface area, concentration, temperature and pressure and catalysts affect the rate of reaction.

Chemicals take part in reactions when their particles collide together. It may be thought of occurring as shown in Figure 14.13.

(i) The effect of surface area

If a chemical is present as a large piece of matter (such as a marble chip) the particles are held together (see Figure 14.14) and the surface of the chip offers a relatively small surface area to the particles of the second reactant (such as hydrochloric acid).

Collisions take place between particles on the surface of the large piece of matter and particles of the second reactant.

If a chemical is present as many small pieces of matter (such as powdered limestone) it has a relatively large surface area. Collisions take place between particles in the smaller pieces of matter and the second reactant, as Figure 14.15 shows.

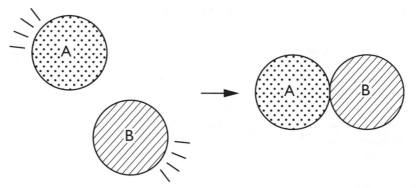

Figure 14.13   A and B particles coming together to form AB particles

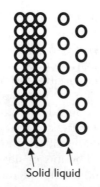

Solid liquid

Figure 14.14   Particles in a solid and particles in a liquid

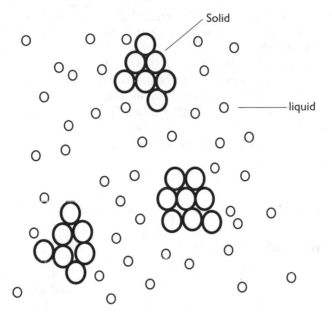

Figure 14.15 Particles in smaller pieces of solid matter and liquid particles

By reducing the size of the piece of matter more particles in it are exposed so that they can collide with the particles of the second reactant and the speed of the reaction is increased.

QUESTION

11 A piece of coal may burn steadily on a fire but coal dust in a mine can cause an explosion. Explain these observations.

(ii) The effect of concentration

The concentration of a reactant is a measure of the number of particles of it that are present in a certain volume. If a reactant is present in a high concentration it has a large number of particles which are available to collide with the second reactant and the reaction is fast. If the reactant is present in a low concentration it has a small number of particles which are available to collide with the second reactant and the reaction is slower.

QUESTION

12 Use the collision theory to explain the results shown in Figure 14.6 (p. 188).

The concentration of a gas is increased by raising its pressure and is decreased by lowering its pressure.

(iii) The effect of temperature

Temperature is a measure of the hotness or coldness of a substance. The hotness or coldness is due to the energy that the particles possess. The particles in a hot substance possess a large amount of energy. This makes them move quickly. The particles in a cold substance possess a smaller amount of energy. This makes them move more slowly.

Faster-moving particles make more collisions than slower-moving particles, just as someone running though a crowd makes more collisions with people than someone who is just walking. As more collisions take place when the particles are moving quickly the speed of the reaction is fast. Fewer collisions take place with slower-moving particles and the reaction is slow.

In most reactions the rate of reaction doubles if the temperature of the reaction is raised by 10°C.

QUESTION

13 Use the collision theory to explain the results shown in Figure 14.9 (p. 191).

(iv) The effect of a catalyst

Many catalysts provide a solid surface on which particles in liquids and gases can react (see Figure 14.16). The different particles are brought into close contact which results in them reacting together. The particles need less energy to react because the catalyst has 'helped' to bring them closer together (see also p. 203). When the particles have reacted they leave the catalyst 's surface and are replaced by more particles which are ready to react.

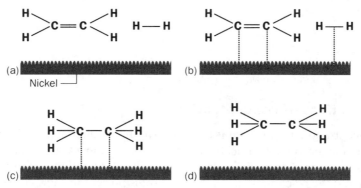

Figure 14.16   The action of a nickel catalyst; the dotted lines show how the molecules are held to the catalyst

QUESTION

14 What would happen if the product particles did not leave the surface of the catalyst?

14.10 A closer look at change

When chemists want to look more closely at a change or chemical reaction they may use the term 'system'. This word is used to describe the reactants, factors and products which are important to the change or reaction and is also used to partition them off from other changes and processes. Biologists use a similar concept when they are studying the way living organisms and environmental factors react in a certain place. The term they use for this is 'ecosystem'. The following are examples of systems which are used to look more closely at changes.

(i) Physical change

A substance undergoes a physical change when it changes from one physical state to another. For example, a physical change occurs when a solid changes into a liquid, a liquid turns into a gas or a gas turns into a liquid.

(a) Non-reversible change

Some physical changes cannot be reversed. For example, if you put a log on a fire it changes into solid ash and carbon dioxide gas and water vapour that escape into the air.

(b) Reversible change

In these changes the substance can change from a solid to a liquid and back again. For example, a piece of wax may be melted and poured into a candle mould then cooled and allowed to set. The change could be described as:

solid	melting	liquid
wax	freezing	wax

The processes of melting and freezing occur at certain temperatures so it is easy to predict the state of individual molecules in the wax.

The processes of evaporation and condensation of water occur over a wide temperature range of 0–100°C, so the state of individual molecules cannot be predicted easily.

QUESTION

> 15 Construct an equation showing how liquid water and water vapour are related through evaporation and condensation.

If a dish of water is left on a bench some of it will evaporate and move away and condense elsewhere. In the dish there will be no reversible reaction taking place and the volume of the water in the dish will decrease until evaporation is complete.

If a flask is half-filled with water then sealed with a stopper the volume of water in the flask will remain constant. However, the two processes of evaporation and condensation have not ceased. They continue as long as the flask remains within the temperature range of 0–100°C. The flask of water forms a closed system in which evaporation and condensation continue. The reason why the volume of liquid water remains the same is that the rate of evaporation and condensation are balanced. The system is said to be **in equilibrium**. As molecules of water evaporate from the surface of the liquid other molecules are condensing on the side of the glass. (This is best seen in a flask kept on a warm windowsill.) The equilibrium is said to be **dynamic equilibrium**. This is a state of equilibrium in which the particles are moving from one state to another but the overall quantities of the states of matter remain unchanged.

(ii) Chemical change

(a) Non-reversible change

Some thermal decompositions are examples of a non-reversible changes. For example, when copper nitrate is heated the following reaction occurs:

Copper nitrate $\rightarrow$ copper oxide + nitrogen dioxide + oxygen
$$2Cu(NO_3)_2 \, (s) \rightarrow 2CuO \, (s) + 4NO_2 \, (g) + O_2 \, (g)$$

The products cannot react directly together to reform the reactants.

(b) Reversible change

Two reactions are involved in reversible chemical reactions. They are the forward reaction (shown by the arrow pointing from left to right in an equation) and the back reaction (shown by the arrow pointing from right to left).

In these reactions the products react directly with each other to form the original reactant or reactants.

(c) Reversible reactions from one reactant

When calcium carbonate is heated it decomposes to form calcium oxide and carbon dioxide but these products can react directly as they cool to form the initial reactant, as the equation shows:

$$\text{Calcium carbonate} \underset{\text{cooling}}{\overset{\text{heating}}{\rightleftharpoons}} \text{calcium oxide + carbon dioxide cooling}$$
$$CaCO_3 \, (s) \rightleftharpoons CaO \, (s) + CO_2 \, (g)$$

In the production of lime (see p. 232) the carbon dioxide must be removed from the kiln to prevent the back reaction.

The reaction can be regarded as a temporary change in the substance due to a break up or dissociation of its components. The word 'dissociation' is used in the description of reversible reactions. The reaction of calcium carbonate just described is an example of thermal dissociation.

(d) Further examples of thermal dissociation

Copper sulphate

When hydrated copper (II) sulphate crystals are heated, water is released and white powder of anhydrous copper (II) sulphate is produced. If water is added to the white powder the hydrated copper (II) sulphate crystals are produced again.

16 Write the symbol equation for the dissociation of hydrated copper (II) sulphate ($CuSO_4 \cdot 5H_2O$).

Ammonium chloride

When ammonium chloride is heated ammonia and hydrogen chloride are produced but as they cool they form ammonium chloride again.

QUESTION

17 Write a symbol equation for the dissociation of ammonium chloride (NH_4Cl).

(e) Reversible reactions with two reactants

Reaction of bismuth (III) chloride

Bismuth (III) chloride forms a colourless solution with concentrated hydrochloric acid but when water is added a white precipitate of bismuth oxychloride forms, as the formula equation shows:

$$BiCl_3 \text{ (aq)} + H_2O \text{ (l)} \rightleftharpoons BiOCl \text{ (s)} + 2HCl \text{ (aq)}$$

The mixture is made colourless again by the addition of more concentrated hydrochloric acid.

QUESTION

18 What would be the effect of adding more water after the extra hydrochloric acid had been added?

Reaction of iodine monochloride

If chlorine gas is passed over iodine a liquid of iodine monochloride forms, as the symbol equation describes:

$$I_2 \text{ (s)} + Cl_2 \text{ (g)} \rightarrow 2ICl \text{ (l)}$$

Iodine monochloride is a brown liquid but if more chlorine is allowed to mix with it a yellow solid of iodine trichloride is formed.

The reaction is described by the following symbol equation:

$$ICl \text{ (l)} + Cl_2 \text{ (g)} \rightleftharpoons ICl_3 \text{ (s)}$$

Figure 14.17 shows the apparatus in which the reaction can take place.

QUESTIONS

19 If chlorine is removed from the system what change will take place in (a) the state of the substance in the U tube, (b) the colour of the substance in the U tube?

20 What would happen if more chlorine was admitted to the tube again?

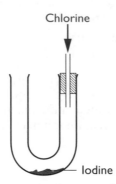

Chlorine

Iodine

Figure 14.17   The reaction between chlorine and iodine

14.11 Chemical equilibrium

On p. 197 the concept of physical equilibrium was described. Here the concept of equilibrium is applied to the reactants and products of a reversible reaction.

A general equation for a reversible reaction is:

$$A + B \rightleftharpoons C + D$$

If the forward reaction is faster than the back reaction the quantities of A and B decline until they are almost used up and the quantities of C and D increase. If the back reaction is faster than the forward reaction the quantities of C and D are almost used up and the amounts of A and B greatly increase.

If the rates of the forward and back reaction are the same the amounts of A, B, C and D remain constant. The actual chemicals from which each one is made change as the reactions take place just as the amount of liquid water and vapour in the sealed flask remained the same but the water molecules moved between the two states of matter.

Potassium chromate is a yellow solution. When sulphuric acid is added to it potassium dichromate forms which is orange The reaction is described in the following equation:

chromate (IV) ions + acid $\rightleftharpoons$ dichromate (VI) ions + water

$$2CrO_4{}^{2-} \ (aq) + 2H^+ \ (aq) \rightleftharpoons Cr_2O_7{}^{2-} \ (aq) + H_2O \ (l)$$

QUESTION

21   What colour would you expect to see if the forward reaction was (a) faster, (b) slower than the backward reaction?

14.12 A closer look at equilibrium

The reactants and products in a general equation:

$$A + B \rightleftharpoons C + D$$

reach equilibrium in the following way. A and B begin to react and produce C and D. As this reaction proceeds, the concentrations of A and B fall and the rate of the forward reaction also begins to fall. As a consequence of the forward reaction the concentrations of C and D increase and they begin to react. This causes the back reaction to begin and to speed up. The dwindling amounts of A and B due to the forward reaction are replenished by the reaction between C and D. Eventually the rates of the two reactions balance each other. When this happens dynamic equilibrium has been reached.

The balance between the two reactions is a delicate one and can be easily upset. In 1888 Henri Le Chatelier described the way the equilibrium could change. This description became known as Le Chatelier's Principle. It can be explained that if the balance of a system in equilibrium is upset the system will react to oppose the effect of the unbalancing factor. It can be applied to the general equation as follows.

(i) Altering the amount of product

The unbalancing effect could be the removal of some of D. If this occurred the back reaction rate would slow down but the forward reaction rate would continue and the concentration of D would be increased. The position of equilibrium is said to be moved to the right of the equation.

(ii) Altering the experimental conditions

If the forward reaction was exothermic and the back reaction was endothermic (see below) raising the temperature of the reactants and products would slow down the forward reaction and speed up the back reaction. This would cause the amounts of A and B to increase. In this case the point of equilibrium is said to move to the left. Le Chatelier's Principle is widely used in industry to find ways of moving the equilibrium position so that large amounts of products can be made as economically as possible. The application of Le Chatelier's Principle is described in the production of ammonia on p. 238 and the production of sulphuric acid on p. 240.

QUESTION

22 What do you think would be the effect on the reaction shown in the general equation by (a) adding more of A and B, (b) adding more of C and D, (c) taking away some of A?

14.13 Reactions and heat

(i) Exothermic reactions

When some reactions take place they release energy in the form of heat. These reactions are called **exothermic reactions**. The most spectacular exothermic reactions are combustion reactions where energy in the form of light also occurs. Burning of any fuel is an example of an exothermic reaction. The heat energy released may be used to cook a meal or warm a home.

You may also feel the heat produced by a chemical reaction produced inside you. This reaction is called respiration and it keeps you alive. When you exercise the rate of respiration increases. Your body produces sweat to take away the heat as evaporation takes place on your skin.

When water is added to anhydrou copper (II) sulphate crystals are formed and heat is given out.

When a neutralisation reaction takes place between an acid and an alkali heat is also given out.

(ii) Endothermic reactions

When some reactions take place they take in heat energy. These reactions are called **endothermic reactions**. In addition to providing heat to speed up the rate of reactions the Bunsen burner is also used to supply heat for decomposition reactions. These reactions are endothermic.

An example of an endothermic reaction is the decomposition of a piece of limestone in a Bunsen burner flame to produce calcium oxide and carbon dioxide. This process is carried out industrially in a lime kiln.

Some endothermic reactions can be detected by a fall in temperature. For example, when potassium chloride is added to water it dissolves and the temperature of the mixture falls.

Photosynthesis is an endothermic reaction which occurs in green plants and produces food for themselves and for animals.

14.14 Energy-level diagrams

The energy change which takes place during a chemical reaction can be represented by an energy-level diagram. Figure 14.18 shows an energy level diagram for an exothermic reaction.

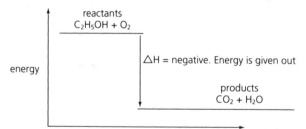

Figure 14.18   Energy-level diagram – exothermic

QUESTION

    23 Construct an energy-level diagram for the combustion of methane (formula methane = CH_4).

Figure 14.19 shows an energy level diagram for an endothermic reaction.

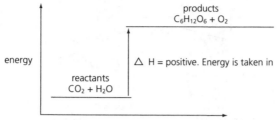

Figure 14.19   Energy-level diagram – endothermic

QUESTION

    24 How are the energy-level diagrams for exothermic and endothermic reactions
        (a) similar, (b) different?

14.15 Activation energy

Many exothermic reactions, particularly combustion, need an input of energy to set them going. For example, a camp fire must be lit with a match to start the reaction but once the reaction has started there is enough energy released to keep it going. This energy which must be given to the reactants before they can react is called **activation energy**. This can be incorporated into a diagram of the energy level for the reaction. When this is done the diagram is known as an **energy profile**. Figure 14.20 shows an energy profile.

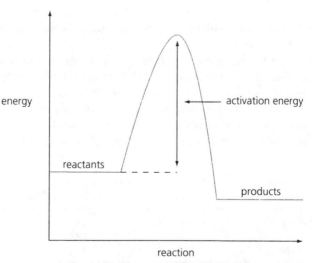

Figure 14.20   Energy profile

    Catalysts reduce the amount of activation energy needed to start a reaction. The effect of a catalyst on the activation energy can be seen by examining an energy profile for a reaction without a catalyst and with a catalyst as shown in Figure 14.21.

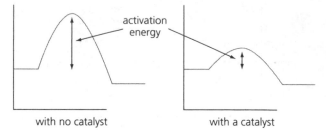

with no catalyst   with a catalyst

Figure 14.21 Effect of catalyst on activation energy

QUESTION

25 Construct an energy profile for the decomposition of hydrogen peroxide showing the effect of the catalyst on the activation energy of the reaction.

14.16 Bonds

The atoms of elements are held together in many compounds by **covalent bonds**. When the bonds are broken energy is taken in and when the bonds form energy is released. During a chemical reaction bonds are broken and new bonds are made. If the energy taken in when the bonds are broken is less than the energy given out when the new bonds are formed there is a surplus of energy and the reaction is exothermic.

If the energy taken in when the bonds are broken is more than the energy given out when the new bonds are made the reaction is endothermic.

The activation energy required by some reactions is needed to break some bonds to start the reaction.

Calculations can be made on the energy changes in reactions by studying a bond energy table and a formula equation of the reaction. For example, Table 14.1 shows the bond energy that exists between two hydrogen atoms, two bromine atoms and an atom of hydrogen and an atom of bromine. The amount of energy is measured in kJ mol⁻¹. The measurement shows the amount of energy required to break the bonds in one mole of the substance possessing the bonds.

Bond	Energy (kJ mol^{-1})
H–H	436
Br–Br	193
H–Br	366

Table 14.1 Breaking bonds

Hydrogen and bromine take part in a reaction in which hydrogen bromide is formed. The symbol equation for this reaction is:

$H_2(g)$	+	Br_2 (l)	$2HBr$ (g)
1 mole		1 mole	2 moles of hydrogen
of hydrogen		of bromine	bromide

The energy needed to break the bonds between the hydrogen atoms in one mole of hydrogen is

$1 \times 436 = 436\,kJ$

The energy needed to break the bonds between the bromine atoms in one mole of bromine is

$1 \times 193 = 193\,kJ$

The total energy taken in to break the bonds of one mole of hydrogen and one mole of bromine taking part in the reaction is

$436 + 193 = 629\,Kj$

The amount of energy released when the bonds form to make two moles of hydrogen bromide is

$2 \times 366 = 732\,kJ$

By comparing the energy taken in by the reactants with the energy released by the product it is seen that $732 - 629 = 103\,kJ$ of energy is surplus to the requirement of the reaction and is released as heat. The calculation shows that the reaction is exothermic.

In work on energy an H symbol is used. This does not represent the element hydrogen. The H symbol is used to show the energy stored in the bonds. This energy is known as **enthalpy**. During a reaction there is a change in enthalpy. This is represented by the symbol ΔH (delta H). It is used in calculations on reactions and in energy-level diagrams. For example, the energy change for the reaction between hydrogen and bromine is $\Delta H = -103\,kJ$. Figure 14.22 shows the energy level diagram for the reaction.

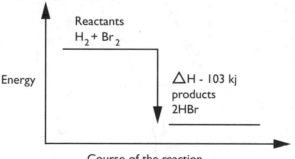

Figure 14.22   Complete energy-level diagram for the reaction

14.17 Fuel

A fuel is a substance which takes part in a combustion reaction in which large amounts of heat energy are released. The first fuel to be widely used was wood. This is still an important fuel in developed countries. Coal, oil and natural gas are fossil fuels which are widely used in developing countries. They are used in power stations to provide the heat to make steam which spins turbines to generate electricity.

14.18 The energy in a fuel

The amount of useable energy in a fuel can be found by measuring the amount of heat energy produced during the combustion of a certain amount of fuel. Figure 14.23 shows the apparatus that can be used to investigate fuels such as butanol, ethanol and propanol.

The mass of the fuel and container is found by weighing on a balance and a mass of 500 g of water is placed in the can. The temperature of the water is taken then the wick of the fuel is lit and the fuel is placed under the can. The water is gently stirred and temperature of the water is checked regularly. When it has risen by 10 degrees the flame is extinguished and the second temperature of the water is taken. This may be a little higher than 10 degrees due to heat in the metal can

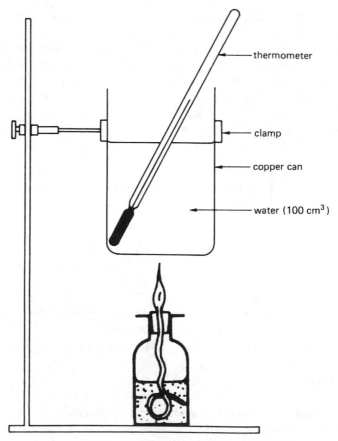

Figure 14.23 Measuring heat released by a fuel

when the flame was extinguished passing into the water. The fuel and container are reweighed.

If the can contained 1 kg of water the water would store 4.2 kJ for every 1°C rise in temperature. As the can contains 0.5 kg the water stored 2.1 kJ for each 1°C rise in temperature. If the water rose by 10°C the amount of energy released by the fuel would be $2.1 \times 10 = 21$ kJ.

The mass of the fuel used is found by subtracting the second mass reading from the first. As the mass is measured in grams the energy in kJ/g is found by dividing the mass of fuel used by 21.

QUESTION

26   How could you find the useable energy stored in a mass of candle wax?

14.19 Summary

- Rates of reaction are affected by:
  (a) particle size (see p. 185)
  (b) concentration (see p. 188)
  (c) temperature (see p. 190)
  (d) catalysts (see p. 190).
- A catalyst speeds up a reaction but does not change chemically (see p. 192).
- Enzymes are biological catalysts (see p. 192).
- Some reactions are affected by light (see p. 193).
- The collision theory can be used to explain how reactions take place (see p. 194).
- Physical changes can be non-reversible or reversible (see p. 197).
- An equilibrium can be reached in a reversible physical change (see p. 197).
- Chemical changes can be non-reversible or reversible (see p. 198).
- An equilibrium can be reached in a reversible reaction (see p. 200).
- An exothermic reaction gives out heat (see p. 201).
- An endothermic reaction takes in heat (see p. 202).
- Activation energy is required to start some reactions (see p. 203).
- Chemicals take in energy when bonds are formed and give out energy when bonds are broken (see p. 204).
- The energy in a fuel can be measured by a simple experiment (see p. 206).

■Ⓥ 15 Organic chemistry

Objectives

When you have completed this chapter you should be able to:
- Describe the properties and uses of **alkanes**
- Explain how **isomers** occur
- Describe the properties and uses of **alkenes**
- Understand how **polymers** are formed
- Describe the properties and uses of **alcohols**
- Describe the preparation of **ethanol**
- Describe the properties and uses of **carboxylic acids**.

The early chemists divided chemicals into two groups – **inorganic** chemicals formed from rocks, water and the air and **organic** chemicals formed by plants and animals. It was believed that organic chemicals could only be made by living things and that a vital or life force was needed to make them. In 1828 Friedrich Wohler showed this idea to be false. He studied the chemical urea which is excreted by animals and discovered a way of making urea from inorganic chemicals.

Today organic chemistry is redefined as the study of compounds made from carbon but excluding carbonates, hydrogen carbonates and carbon dioxide. These compounds are inorganic compounds. The main sources of organic chemicals are the fossil fuels.

15.1 Organic compounds

There are millions of organic compounds. Most of the atoms in these compounds are joined together by **covalent bonds**. The compounds are divided into groups according to their molecular structure. The simplest organic compounds are called alkanes.

(i) Alkanes

In alkanes all the carbon atoms are linked together by a single covalent bond (one shared pair of electrons). The only other element in an alkane molecule is hydrogen. Molecules made of carbon and hydrogen only are called **hydrocarbons**.

The simplest alkane is methane. It has just one atom of carbon which forms covalent bonds with four atoms of hydrogen. Organic compounds with two carbon atoms have eth- the first part of their name. Other first parts of names are prop- (three carbon atoms) but- (four carbon atoms), pent- (five carbon atoms) and hex- (six carbon atoms).

(a) The molecular structure

The six simplest alkanes are shown in Figure 15.1.

It can be seen that when the molecules in a group are arranged in order of size

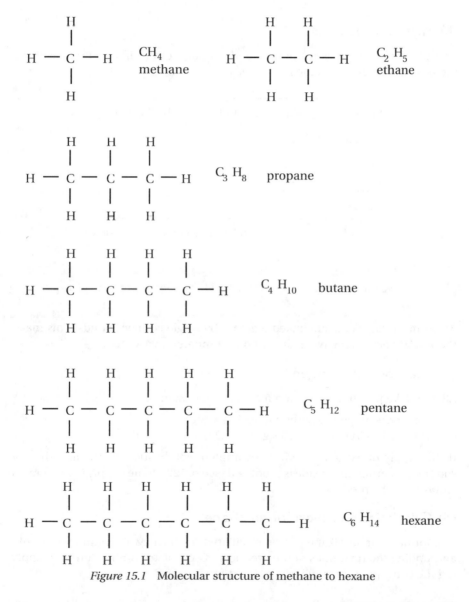

Figure 15.1 Molecular structure of methane to hexane

each one differs from the next by one carbon atom and two hydrogen atoms. These have a relative molecular mass of 14. A series of molecules which differ by a certain relative molecular mass is called an **homologous series**. Many organic compounds form a homologous series like the alkanes.

The molecules in a homologous series can be represented by a general formula. If the molecular formulae of the first six alkanes are examined it can be seen that they all fit the general formula $C_nH_{2n} + 2$.

QUESTION

  1 What is the molecular formula of an alkane with (a) 8, (b) 18 carbon atoms? Draw a diagram for each showing the arrangement of all the atoms present.

(b) Physical properties

The physical properties of melting and boiling points show a trend with the increase in size of the alkane molecule, as shown in Table 15.1.

Alkane	Melting point (°C)	Boiling point (°C)
methane	−183	−162
ethane	−172	−89
propane	−187	−42
butane	−135	−0.5
pentane	−130	36
hexane	−95	69

Table 15.1 The melting and boiling points of six alkanes

QUESTION

  2 Which alkanes are (a) gases, (b) liquids at room temperature? Explain your answer.

All atoms in an alkane are linked together by single covalent bonds. This makes them relatively unreactive hydrocarbons compared to alkenes.

(c) Reaction with oxygen

Alkanes take part in combustion reactions. For example:

  ethane + oxygen → carbon dioxide + water
  $2C_2H_6$ (g) + $7O_2$ (g) → $4CO_2$ (g) + $6H_2O$ (l)

If the supply of oxygen is limited incomplete combustion occurs and the poisonous gas carbon monoxide is produced (see p. 120). If the supply is very limited carbon may be produced.

(d) Calculating a formula for an alkane

The formula for an alkane can be calculated by burning a certain amount of it and finding the quantities of the products of complete combustion then applying the concept of the mole (see p. 161) to the results.

EXAMPLE

20 cm³ of an alkane CxHy produces 60 cm³ of carbon dioxide and 80 cm³ of water vapour.

The formula for the alkane can be found by comparing the ratio of the quantities as follows:

(1) 20 cm³ of alkane CxHy produces 60 cm³ of carbon dioxide
This can be restated as

I volume of alkane produces 3 volumes of carbon dioxide

and by considering moles (see p. 161) it can be revised to

I mole of alkane produces 3 moles of carbon dioxide

This leads to the statement that

I molecule of alkane produces 3 molecules of carbon dioxide

From this the alkane molecule can be seen to contain 3 atoms of carbon

$x = 3$

(2) 20 cm³ of alkane C_3Hy produces 80 cm³ of water vapour
This can be restated as

I volume of alkane produces 4 volumes of water

or

I mole of alkane produces 4 moles of water

or

I molecule of alkane produces 4 molecules of water

As there are two atoms of hydrogen in each molecule of water there must be $2 \times 4 = 8$ atoms of hydrogen in the alkane.

$y = 8$

From (1) and (2), the formula of the alkane is C_3H_8.

QUESTION

3 What is the formula of a hydrocarbon which took part in the following reaction? When 30 cm³ of the alkane was burnt in air 180 cm³ of carbon dioxide and 210 cm³ of water vapour was produced.

(ii) Saturated compounds

A saturated compound is one in which all the atoms are linked together by single covalent bonds. Saturated compounds take part in substitution reactions as described in the next section.

(a) Substitution reaction

In a substitution reaction one of the hydrogen atoms in a molecule is replaced by the atom of another element or by a group of atoms.

When methane is mixed with chlorine and the mixture is exposed to sunlight one of the hydrogen atoms in each molecule is substituted by a chlorine atom, as the following equations show:

methane + chlorine → chloromethane + hydrogen chloride

CH_4 (g) + Cl_2 (g) → CH_3 Cl (g) + HCl (g)

Bromine and alkanes also take part in substitution reactions. However alkanes do not react with bromine water (see p. 216).

(iii) Isomers

The atoms in a molecule such as butane can be arranged in two ways. Figure 15.2 shows a second way in which the atoms of butane in Figure 15.1 can be arranged.

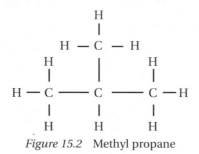

Figure 15.2 Methyl propane

If you count the number of carbon and hydrogen atoms in the butane molecule in each diagram you will find that they have the same molecular formula, $C_4 H_{10}$. Compounds which have the same molecular formula but different molecular structures are called **isomers**.

The feature which distinguishes the two isomers of butane is the side branch on the second atom from the left in Figure 15.2.

The prefix of meth- and prop- are used in the naming of the isomer. Meth- refers to the short branch which is like a methane molecule and prop- refers to the longer branch which is like a propane molecule. Meth- and prop- are brought together to form the name methyl propane.

QUESTION

4 Which part of the methyl propane molecule does the (a) methyl part, (b) propane part refer?

Pentane $C_5 H_{12}$ has three isomers. One is shown in Figure 15.1 and a second, with two side branches on a carbon atom at the end of a short chain of three carbon atoms, is shown in Figure 15.3.

QUESTIONS

5 Draw the third isomer of pentane. It has two side branches attached to the central carbon atom in the short chain of three carbon atoms.
6 Draw the isomers of hexane.

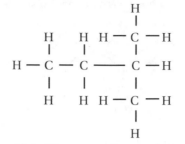

Figure 15.3 The second isomer of pentane

The isomer of an alkane without side branches has strong intermolecular forces which hold the molecules together. Molecules with long chains of carbon atoms have stronger intermolecular forces than molecules with shorter chains. The presence of side branches on molecules reduces the strength of the intermolecular forces between them. Molecules with a large number of side branches have weaker forces holding them together than molecules with a smaller number of branches. The strength of the intermolecular forces affects the physical properties, such as the melting and boiling points, of the isomer.

QUESTION

7 Butane has a melting point of −138°C and a boiling point of 0°C. Methyl propanone has a melting point of −159°C and a boiling point of −12°C. Why is there a difference in these properties?

15.2 Uses of alkanes

As alkanes release large amounts of energy in combustion reactions they are used as fuels. Alkanes with small molecules such as methane are used in their gaseous form. They are colourless and odourless and have chemicals added to them to give them a smell. Larger molecules form liquids. For example, octane is used in petrol. The largest molecules form waxy solids and are used to make candles. Alkanes form the group sometimes called paraffins.

QUESTION

8 Why do you think odourless gaseous fuels are given chemicals to make them smell?

15.3 Alkenes

Alkenes are hydrocarbons which have a pair of carbon atoms linked together by a double covalent bond. The carbon atoms share two pairs of electrons instead of one pair as occurs in alkanes.

The simplest alkene is ethene. It has a molecular structure shown in Figure 15.4.

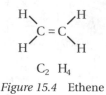

$C_2 H_4$

Figure 15.4   Ethene

QUESTION

9 Why does the alkene methene not exist?

Propene and butene are two further examples of alkenes and their molecular structures are shown in Figure 15.5.

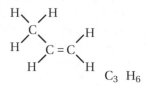

$C_3 H_6$

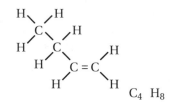

$C_4 H_8$

Figure 15.5   Structure of propene and butene

If you count the number of carbon atoms and hydrogen atoms in each molecule you will find that they can be represented by the general formula C_nH_{2n}.

Alkenes form a homologous series like alkanes do (see p. 210). Table 15. 2 shows the melting points and boiling points of three alkenes.

Alkene	Melting point (°C)	Boiling point (°C)
ethene	−169	−102
propene	−185	−48
butene	−185	−7

Table 15.2   Melting and boiling points of alkenes

QUESTION

10 Look at Tables 15.1 and 15.2 and compare the melting points and boiling points of alkanes and alkenes. How are they (a) similar, (b) different?

(i) Preparation of alkenes

Alkenes can be prepared in two ways.

(a) The dehydration reaction

In this reaction a water molecule is removed from an alcohol molecule.

For example in the dehydration of ethanol ethene is produced. The reaction can take place by:

(1) heating a mixture of ethanol and concentrated sulphuric acid. The acid is present in excess to make sure the dehydration process is complete.
(2) heating ethanol to form a vapour which is then passed over hot aluminium oxide. The aluminium oxide acts as a catalyst.

The formula equation for the dehydration reaction is:

$$C_2H_5OH \text{ (l)} \rightarrow C_2H_4 \text{ (g)} + H_2O \text{ (l)}$$

(b) Cracking

Alkenes can be produced from alkanes by a process called cracking (see p. 233).

15.4 General properties of ethene

Ethene is a colourless, neutral gas which is slightly less dense than air and has a slightly sweet smell. It is insoluble in water.

15.5 Ethene and oxygen

Pure ethene will not support a burning taper but when it is mixed with oxygen a combustion reaction occurs. If the amount of oxygen is limited carbon or carbon monoxide is produced but if there is a plentiful supply of oxygen carbon dioxide and water is produced. The symbol equation for the reaction is:

$$C_2H_4 \text{ (g)} + 3O_2 \text{ (g)} \rightarrow 2CO_2 \text{ (g)} + 2H_2O \text{ (g)}$$

15.6 Unsaturated compounds

An unsaturated compound is one in which two of the atoms present in the molecule share a double bond. The compound takes part in addition reactions which are described in the next section.

QUESTION

11 How are saturated and unsaturated compounds different (see also p. 211 before you answer).

15.7 Addition reactions

(i) With bromine water

The addition reaction which is used to test for the double bond of alkenes is the test with bromine water. In this test the suspected alkene is shaken with bromine water. If the alkene is present the bromine water changes from brown to colourless. The symbol equation for the reaction of ethene and bromine is:

$$C_2H_4 \text{ (g)} + Br_2 \text{ (aq)} \rightarrow C_2H_4Br_2$$

This equation can also be written to show the structure of the molecules, as shown in Figure 15.6.

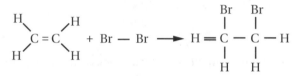

Figure 15.6   Structural formulae of reactants and product

(ii) With water

Ethene reacts with water in the form of steam to form **ethanol**. The reaction takes place at a temperature of 300°C and a pressure of 60 atmospheres. It is speeded up by the use of silica pellets which contain phosphoric (V) acid which act as a catalyst. The symbol equation for the reaction is:

$$C_2H_4 \text{ (g)} + H_2O \text{ (g)} \rightarrow C_2H_5OH \text{ (g)}$$

Most industrial ethanol is prepared in this way.

(iii) With hydrogen

Ethene reacts with hydrogen to form ethane. The symbol equation for the reactions is:

$$C_2H_4 \text{ (g)} + H_2 \text{ (g)} \rightarrow C_2H_6 \text{ (g)}$$

The reaction is aided by a heated nickel catalyst.

(iv) Using hydrogen to make margarine

Double bonds are found between pairs of carbon atoms in molecules of vegetable oils. The oils are described as **unsaturated**. They are saturated by an addition reaction with hydrogen. The oil is vapourised, mixed with hydrogen and passed over a heated nickel catalyst. The product is more fully saturated molecules which make a semi-solid fat. This is used to make margarine.

(v) Making polymers

If ethene molecules are heated under high pressure in the presence of a catalyst an addition reaction takes place in which the ethene molecules are linked together. Figure 15.7 shows the change in structure of the molecules to form a molecule with a long chain of carbon atoms called **polyethene**. This substance is widely known as **polythene**.

The equations for the reaction are:

ethene → polyethene
$$n(H_2C = CH_2) \ (g) \rightarrow -(-CH_2-CH_2-)-n \ (s)$$

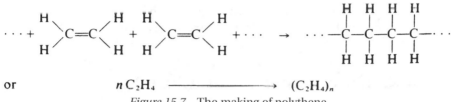

or $\qquad n \, C_2H_4 \xrightarrow{\hspace{3cm}} (C_2H_4)_n$

Figure 15.7 The making of polythene

QUESTION

12 Styrene is a liquid alkene that has the structural formula $C_6H_5 - CH = CH_2$. (a) draw the structure of the molecule, (b) write the equation in which it forms the solid polystyrene.

The small molecules which link together to form long chains are called **monomers**. The long chains which are produced when the monomers link together are called **polymers**.

15.8 Uses of ethene

When alkenes burn they make a sooty flame and are not used as fuels. Ethene is used to make polymers, ethanol and a compound called ethyl glycol, commonly known as **antifreeze**, which is added to car engine water systems in winter.

15.9 Alcohols

Ethanol, the alcohol in alcoholic drinks, is just one of a homologous series of alcohols. Each alcohol molecule has a hydroxe group –OH. This is called the functional group which give the alcohols their properties. The –OH group is not ionic so it does not make the alcohols alkaline.

The general formula for alcohols is $C_n H_{2n+1} OH$.

Figure 15.8 shows the structure of the first three alcohols in the homologous series.

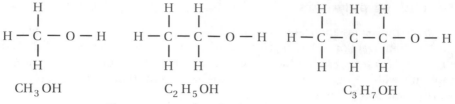

CH_3OH C_2H_5OH C_3H_7OH

Figure 15.8 Methanol, ethanol and propan1ol

QUESTIONS

13   Which alcohol in Figure 15.8 has an isomer? (See also p. 212 to help you.)
14   What do you think the 1 refers to in propan1ol?
15   Construct the molecules for butan1ol and pentan1ol.

15.10 General properties of ethanol

Ethanol is a colourless, neutral liquid which mixes with water.
It has a boiling point of 78°C.

(i) Ethanol preparation

Ethanol has been produced for thousands of years by the process of **fermentation**. In this process some of the enzymes produced by yeast act as catalysts in the break down of sugar into ethanol and carbon dioxide. The symbol equation for the break down of glucose by fermentation is:

$$C_6H_{12}O_6 \text{ (aq)} \rightarrow 2C_2H_5OH \text{ (aq)} + 2CO_2 \text{ (g)}$$

Fermentation is a batch process. This means that containers of sugar and yeast are set up (the batch) and fermentation is allowed to continue until the ethanol has been produced. As the catalyst is an enzyme the reactions must take place below about 45°C. The low temperature ensures that the reaction must be slow. As the source of sugar (sugar beet or sugar cane) is not pure the ethanol that is produced is also not pure. It is purified by fractional distillation.

Ethanol can also be made from ethene. This is a continuous process. The reactants are added to one part of the chemical plant in which the ethanol is made and the pure product is removed from another part of the plant. The ethene is made from hydrocarbons which come from oil (see p. 216).

QUESTIONS

16   Compare the two methods of making alcohol for speed, purity and type of resource (i.e. renewable or non-renewable).
17   The fermentation process takes place in bread making. (a) Which product do you think makes the bread rise. Explain your answer (b) Ethanol is not present in the baked bread. Why?

(ii) Reaction with oxygen

If the amount of oxygen is limited carbon monoxide and carbon may be produced in the incomplete combustion. If the complete combustion of ethanol takes place carbon dioxide and water are produced.

QUESTION

18 Construct a symbol equation for the complete combustion of ethanol. (One molecule of ethanol combines with three molecules of oxygen.)

If a bottle of wine is left open the ethanol in it reacts with oxygen in the air and certain microorganisms present in the liquid. In this oxidation process the ethanol becomes ethanoic acid which makes the wine sour. The chemical equation for the reaction is:

$$C_2H_5OH \ (l) + 2[O] \ (g) \rightarrow CH_3CO_2H \ (l) + H_2O \ (l)$$

This oxidation process is slow. It can be speeded up by the use of catalysts which have been acidified.

When potassium dichromate is used the mixture changes from orange to green. When potassium manganate (VII) is used the mixture changes from purple to colourless.

(iii) Reaction with sodium

When sodium reacts with an alcohol it dissolves in it. This is represented by the state symbol (alc). Hydrogen is produced which is released as a gas.

The symbol equation for the reaction between methanol and sodium is:

$$2CH_3OH \ (l) + 2Na \ (s) \rightarrow 2CH_3ONa \ (alc) + H_2 \ (g)$$

(iv) Reaction with an organic acid

When an alcohol reacts with an organic acid such as ethanoic acid a compound called an **ester** is formed. Esters are a group of compounds which give scents to flowers and flavours to fruits. The symbol equation for the reaction between methanol and ethanoic acid is shown below. The ester formed in this reaction is called methyl ethanoate:

$$CH_3OH \ (l) + CH_3CO_2H \ (l) \rightleftharpoons CH_3CO_2CH_3 \ (l) + H_2O \ (l)$$

15.11 Uses of ethanol

Ethanol is used to make alcoholic drinks such as beer (about 4% alcohol), wine (about 11% alcohol) and whiskey (about 40% alcohol). Ethanol is used as a solvent in the making of perfumes, antiseptics and polishes. In countries such as Brazil it is used as a fuel and mixed with petrol. Ethanol is also used as fuel in camping stoves in the form of **methylated spirits**. Methanol, which is poisonous, is added to the ethanol in methylated spirits to prevent people using the fuel as an alcoholic drink.

19 The melting points and boiling points of methanol and ethanol are −97°C and 65°C, −114°C and 78°C, respectively. (a) Arrange this information in a table, (b) add an extra line to the table and insert the word 'propanol' then select its melting and boiling points from these three pairs: (i) −126°C and 97°C, (ii) −102°C and 70°C, (iii) −180°C and 150°C. (c) Explain your answer to (b).

15.12 Carboxylic acids

The carboxylic acids form a homologous series. Figure 15.9 shows the molecular structure of the first three acids in the series.

If you count the carbon, hydrogen and oxygen atoms in each molecule you will see that the general formula for carboxylic acids is $C_nH_{2n+1}CO_2H$. The functional group which gives the compounds their properties is the carboxyl group $-CO_2H$.

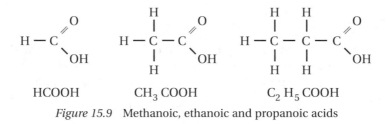

HCOOH $CH_3\,COOH$ $C_2\,H_5\,COOH$

Figure 15.9   Methanoic, ethanoic and propanoic acids

(i) Properties of carboxylic acids

The simplest acid – methanoic acid – is responsible for the stinging sensation that is felt when a nettle leaf is touched. All the acids dissociate to produce H^+ ions but they do not dissociate as strongly as mineral acids. The pH of carboxylic acids range from 6 down to 3.5. Despite their comparative weakness carboxylic acids react with other elements and compounds in a way that is typical of acids.

(ii) Reaction with a metal

When the acids react with a metal, hydrogen is produced.

EXAMPLE

ethanoic acid + magnesium → magnesium ethanoate + hydrogen

$2C_2H_5CO_2H$ (aq) + Mg (s) → $(CH_3CO_2)_2Mg$ (aq) + H_2 (g)

(iii) Reaction with a carbonate

When the acids react with carbonates carbon dioxide is produced.

EXAMPLE

> ethanoic acid + magnesium carbonate → magnesium ethanoate + carbon dioxide + water
>
> $2C_2H_5CO_3H$ (aq) + $MgCO_3$ (s) → $(CH_3CO_2)_2Mg$ (aq) + CO_2 (g) + H_2O (l)

(iv) Reaction with an alkali

A neutralisation reaction occurs between a carboxylic acid and an alkali.

EXAMPLE

> ethanoic acid + sodium hydroxide → sodium ethanoate + water
>
> $C_2H_5CO_2H$ (aq) + NaOH (aq) → $C_2H_5CO_2Na$ (aq) + H_2O (l)

The carboxylic acids react with alcohols to form esters (see p. 219).

15.13 Uses of ethanoic acid

A 3% solution of ethanoic acid forms vinegar which is used as a food preservative. As ethanoic acid is weaker than mineral acids it can be used as a dilute solution to remove scale and furring in kettles without attacking the metal in the kettle. Pure ethanoic acid is used as a solvent and as a raw material for making some polymers.

15.14 Summary

- Alkanes are hydrocarbons which react with oxygen and halogens (see pp. 208–12).
- Isomers are compounds with the same molecular formula but different molecular structures (see p. 212).
- Alkenes are hydrocarbons which have a double bond (see p. 213).
- The double bond can be detected in a test using bromine water (see p. 216).
- Alkenes react with water and hydrogen (see p. 216).
- Alkenes are used to make polymers (see p. 217).
- Alcohols have an – OH functional group (see p. 217).
- Ethanol is made by fermentation or from ethene and steam (see p. 218).
- Alcohols react with carboxylic acids to form esters (see p. 219).
- Carboxylic acids have a functional group – CO_2H (see p. 220).
- Carboxylic acids react with metals, carbonates and alkalis (see p. 220).

■ ᴍ̌ 16 Chemistry in industry

Objectives

When you have completed this chapter you should be able to:
- Describe the processes in the **extraction of iron**
- Describe how **aluminium is extracted**
- Explain how **copper is purified**
- Describe the **fractional distillation of oil**
- Distinguish between **iron** and **steel**
- Distinguish between different **alloys**
- Understand how **metal working** changes the properties of metals
- Describe the production and uses of **lime**
- Describe how **ceramics** and **glass** are made
- Understand the purpose of the **cracking process** in the oil industry
- Distinguish between the different types of **polymer**
- Describe how **plastics react to heat**
- Describe the production and action of **soap**
- Explain the properties of a **composite material**
- Describe the manufacture and uses of **ammonia**
- Describe the manufacture and uses of **nitric acid**
- Describe the manufacture and uses of **sulphuric acid**
- Describe the **special properties of sulphuric acid**.

The first people used natural materials for all their needs. Rocks were used for building and flint nodules were broken up to make knives and axes. Wood was also used for building and for making handles of tools and as a fuel. Wool and skins were used for clothing. Bone and shells and native metals were used for decorative purposes. In time, it was discovered how to make pottery and glass and how to extract metals from their ores. In the late nineteenth century the value of oil (petroleum) as a natural resource was realised. Today we can see the application of chemistry in every part of industry supplying us with the products we need for our daily lives.

This chapter is divided into two sections – extracting materials and making materials.

16.1 Extraction of materials

(i) Metals

The discovery and the use of metals follows the reverse order of the metals in the reactivity series (see p. 108). The most unreactive metals, gold and silver, were discovered first as they could be found uncombined with other elements in their natural or native state. Some native copper can also be found.

Other metals form compounds in rocks. A rock which is particularly rich in a metal is called an **ore**. The chance placing of rocks which were metal ores around a camp fire may have lead to a reduction by carbon in the wood to release the metal as an element. The first metals released in this way were copper and tin.

Metals can be mixed. They form mixtures called **alloys**. When copper and tin are mixed together by melting them they form an alloy called bronze. It is stronger than the two metals from which it is made. Bronze was given such a wide range of uses that the artefacts such as swords, helmets and jewellery from that time led to the time being called the Bronze Age.

When bellows were used on charcoal higher temperatures were obtained which allowed carbon to reduce higher metals in the reactivity series such as lead and iron. When iron was discovered its physical properties, such as hardness, were found to be more useful than bronze. Iron replaced bronze and gave rise to another archeological age – the Iron Age.

Carbon is about midway in the reactivity series. It cannot be used to reduce metals above it in the series. With the development of electrolysis it became possible to extract these metals.

When a metal is reduced to extract it from its ore an oxidation also occurs. This process in which both an oxidation and a reduction takes place is called a **redox reaction**.

(ii) Extracting iron

There are four main types of iron ore. They are haematite containing iron (II) oxide (FeO), magnetite containing iron (III) oxide (Fe_2O_3), pyrite containing iron sulphide (FeS) and spathic ore containing iron carbonate ($FeCO_3$).

Iron is extracted from its ore in a blast furnace (see Figure 16.1).

Iron, coke and limestone are released into the top of the blast furnace. Jets of hot air enter the blast furnace near its base and heat the coke (the source of carbon) to about 2000°C. At this temperature the carbon in the coke is oxidised to carbon monoxide, as the symbol equation shows:

$$2C\,(s) + O_2\,(g) \rightarrow 2CO\,(g)$$

The carbon monoxide is the reducing agent in the extraction of iron. The symbol equation for this reaction is:

$$3CO\,(g) + Fe_2O_3\,(s) \rightarrow 2Fe\,(l) + 3CO_2$$

The temperature of the blast furnace is above the melting point of iron so the metal flows down and collects at the furnace's base.

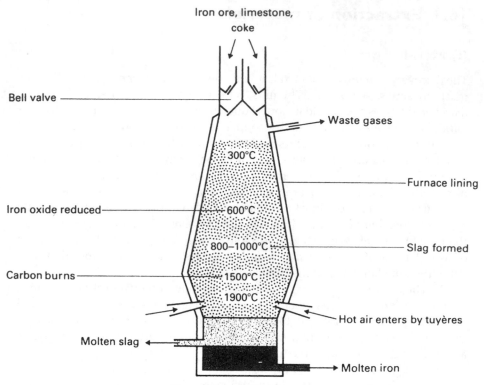

Iron ore, limestone, coke

Bell valve

Waste gases

300°C

Furnace lining

Iron oxide reduced — 600°C

800–1000°C — Slag formed

Carbon burns — 1500°C

1900°C

Hot air enters by tuyères

Molten slag

Molten iron

Figure 16.1 A blast furnace

Iron contains silica. This is separated from the metal by the use of limestone. The temperature in the middle part of the furnace is about 1000°C. At this place the limestone decomposes, as the symbol equation shows:

$$CaCO_3\ (s) \rightarrow CaO\ (s) + CO_2\ (g)$$

The carbon dioxide escapes as a waste gas but the calcium oxide combines with the silica in the ore to form calcium silicate, as the symbol equation shows:

$$CaO\ (s) + SiO_2\ (s) \rightarrow CaSiO_3\ (s)$$

The high temperature of the blast furnace is above the melting point of calcium silicate and it flows to the furnace's base.

QUESTION

1 What are the word equations for the reactions that take place inside a blast furnace?

Calcium silicate is known as slag. It is less dense than the molten iron and floats on top of it. The two molten substances are removed from the blast furnace separately.

When slag cools and solidifies it is used for making the foundations in the construction of roads and buildings. The iron released from the blast furnace is called 'pig iron'. It gets its name from the way it used to be run out into channels. There was a main channel with a number of side channels, and they looked like piglets laying next to their mother. The pig iron is treated to make cast iron, wrought iron and steel (see Making materials, p. 228). The waste gases leaving the top of the blast furnace are used to heat the air which is blown into the pipes near the furnace's base.

QUESTION

2   Make a flow diagram of the extraction of iron showing all the materials used and produced in the process.

(iii) Extracting aluminium

There is just one ore from which aluminium is extracted. It is called **bauxite**. The aluminium compound in the ore is hydrated aluminium oxide. The ore also contains iron oxide which makes the bauxite an impure source of aluminium.

The bauxite is treated with a hot strong solution of sodium hydroxide. Aluminium oxide dissolves in this solution but iron oxide does not. The two oxides are separated by filtering. The aluminium oxide is crystallised from the filtrate then dried to remove the water of crystallisation.

The aluminium oxide is then dissolved in molten sodium aluminium fluorite (cryolite) and is poured into a cell like the one shown in Figure 16.2.

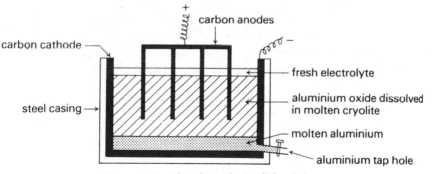

Figure 16.2   The electrolysis of aluminium

QUESTION

3   The melting point of aluminium oxide is 2072°C. The temperature of the molten cryolite in which the aluminium oxide dissolves is 970°C. How does the use of cryolite make the extraction process more economical?

The half equation for the reactions occurring at the electrodes are:

(a) At the cathode

$$4Al^{3+} (l) + 12e^- \rightarrow 4Al (l)$$

(b) At the anode

$$6O^{2-} (l) - 12e^- \rightarrow 3O_2 (g)$$

QUESTION

   4 Where does oxidation and reduction take place in the electrolytic cell?

When the two half equations are combined the full symbol equation for the reaction is:

$$2Al_2O_3 (l) \rightarrow 4Al (l) + 3O_2 (g)$$

The molten aluminium, which is 99.9% pure, collects at the bottom of the cell and is removed regularly while more electrolyte is added to the top. Some oxygen is removed from the cell but a reaction takes place between oxygen and carbon at the anode, as the following symbol equation shows:

$$2C (s) + O_2 (g) \rightarrow 2CO (g)$$

QUESTION

   5 How will the reaction between oxygen and carbon affect the carbon anodes?

(iv) Copper

The main copper ore is copper pyrites. This is a compound of copper iron and sulphur. The copper is separated from the other elements in the compound by heating the copper pyrites with a limited amount of air. The copper that is produced contains bubbles of sulphur dioxide and is called blister copper. It also has other metals in it as impurities.

   Copper is purified by electrolysis using a cell like the one shown in Figure 16.3.

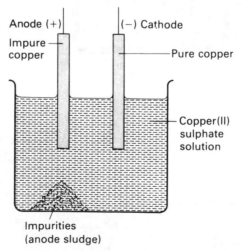

Figure 16.3   Electrolysis used to purify copper

The blister copper is set up as the anode and a thin sheet of pure copper is set up as the cathode. When electricity is passed through the circuit the copper at the anode dissolves in the electrolyte and the copper ions in the electrolyte receive electrons at the cathode and form a metal block which is 99.92% pure.

The half equations which takes place at the electrodes are:

(a) At the anode

$$Cu\,(s) - 2e^- \rightarrow Cu^{2+}\,(aq)$$

(b) At the cathode

$$Cu^{2+}\,(aq) + 2e^- \rightarrow Cu\,(s)$$

QUESTION

    6 Where does oxidation and reaction take place in the purification of copper?

As the copper at the anode dissolves in the electrolyte the metal impurities which do not dissolve fall to the bottom of the cell and form anode sludge. This material is very valuable as it contains metals such as gold, silver and platinum.

(v) Reducing one metal with another

Titanium in the form of titanium chloride is reduced by sodium or magnesium to be released in its metallic form. The symbol equation for the reaction with magnesium is:

$$TiCl_4 + 2Mg\,(l) \rightarrow Ti\,(s) + 2MgCl_2\,(l)$$

QUESTION

    7 Produce a symbol equation for the reaction with sodium. (Note that the solid NaCl is produced.)

Titanium is a versatile metal. It is used as a white pigment in paints and its strong and corrosion resistant properties make it a useful addition to alloys. They are used in a wide variety of ways from making hip replacements to spacecraft.

(vi) Extracting oil products

Oil contains a huge range of organic chemicals. The number of the carbon atoms in their molecules ranges from one to over 50. The chemicals are separated by **fractional distillation**.

In fractional distillation of oil a tall tower is used which has trays inside it. Each tray has tubes called risers which pass through them. Each riser has a top called a bubble cap. When the hot oil is introduced into the tower its vapours rise up through the trays and condense at different levels. They do this because the vapours have different boiling points and as they rise through the tower they cool and condense. The vapours with the highest boiling points condense in the lower

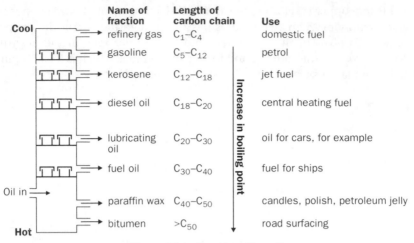

Name of fraction	Length of carbon chain	Use
refinery gas	C_1-C_4	domestic fuel
gasoline	C_5-C_{12}	petrol
kerosene	$C_{12}-C_{18}$	jet fuel
diesel oil	$C_{18}-C_{20}$	central heating fuel
lubricating oil	$C_{20}-C_{30}$	oil for cars, for example
fuel oil	$C_{30}-C_{40}$	fuel for ships
paraffin wax	$C_{40}-C_{50}$	candles, polish, petroleum jelly
bitumen	$>C_{50}$	road surfacing

Figure 16.4 Fractionating oil

regions of the tower while the vapours with the lower boiling points condense further up. Some compounds with short chains such as methane and butane have such low boiling points that they do not condense and are collected and stored in their gaseous forms.

The liquids which have condensed at different levels in the tower and the solids which have formed at the bottom are called **fractions**. Each fraction contains a mixture of chemicals with carbon chains of fairly similar lengths. They are drawn out of the tower and may be refined by further fractional distillation or used in other processes (see cracking, p. 233).

Figure 16.4 shows a fractionating column and some of the uses of the fractions.

QUESTION

8 Which of the oil products in Figure 16.4 have you used today?

16.2 Making materials

(i) Metals

(a) Iron

Cast iron

Some of the pig iron from the blast furnace is slightly refined to produce an iron with a 5% carbon content. This metal is very runny when it is molten and can be used in moulds to make complicated shapes. When the metal cools it is strong, corrosion-resistant but brittle. It is used for making engine blocks for cars and trucks and manhole covers.

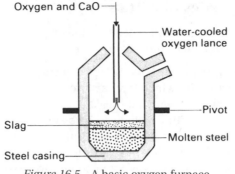

Figure 16.5 A basic oxygen furnace

Wrought iron

This is very pure iron. It is produced by removing all the carbon from pig iron. Wrought iron is not brittle like cast iron and can be bent to make ornamental shapes in ironwork. It can also be used for making iron fences and gates.

(b) Steel

Steel is made by blowing oxygen across the surface of molten pig iron. This process takes place in a basic oxygen furnace. The oxygen enters the furnace in a water-cooled pipe. It removes carbon from the metal (see Figure 16.5).

The carbon is oxidised to carbon monoxide and carbon dioxide. Sulphur is also present in the iron. It is oxidised to sulphur dioxide and leaves the top of the furnace with the other gases. The oxygen also converts silicon in the molten metal to silca and phosphorus to phosphorus (V) oxide. Lime is added to the furnace to mix with these chemicals to make slag. The furnace does not need a fuel like the blast furnace because the sufficient heat is generated by the exothermic reactions taking place.

(c) Types of steel

The amount of carbon remaining in the metal is controlled by the time for which the oxygen blows across its surface. Different mixtures of carbon and iron can be produced. Each one is an alloy of a metal and a non metal. The alloy is called steel. The main types of carbon steel are:

- High carbon steel or hard steel with a carbon content of 0.5–1.5%. It is tough and brittle and used for making knives, razor blades and scissors.
- Medium carbon steel with a carbon content of 0.25–0.5%. It is used for making rails and springs.
- Mild steel with a carbon content of 0.1–0.25%. It is used for girders and for steel plates for ships.

QUESTION

9 How does the amount of carbon mixed with iron affect the hardness of the alloy?

(ii) Useful properties of metals

The properties of metallic elements are described on p. 107. The properties which affect the usefulness of a metal are its:

- **hardness** – the ease with which its surface deforms
- **toughness** – the ease with which it resists breaking or snapping
- **brittleness** – the ease with which it breaks
- **malleability** – the ease with which it can be beaten or rolled into shape
- **ductility** – the ease with which it can be pulled into a wire
- **tensile strength** – the strength it possesses to resist a pulling force
- **compressive strength** – the strength it possesses to resist squashing
- **resistance to corrosion**
- **low density** (lightness in weight).

QUESTIONS

10 How do you think the properties of hardness, malleability and ductility are related?

11 Can a metal be (a) hard and brittle, (b) tough and brittle? Explain your answers.

(iii) Alloys

Different metals can be mixed together in a molten state. When they cool they form a solid metal which has different properties from the metals which formed the mixture. By controlling the amounts of different metals in a mixture an alloy can be made which has properties that make it suitable for a particular task as the following examples show:

- **Brass** is a mixture of about 60% copper and 40% zinc. It is a good conductor of electricity and is used to make electrical contacts. It is corrosion-resistant which makes the contacts last a long time. Brass has a shiny surface and an attractive colour and is easily shaped. These properties make it useful for door handles and ornaments.
- **Bronze** is a mixture of 90% copper and 10% tin. It is a strong, hard, corrosion-resistant alloy which is easy to cast into complicated shapes. It is used for machine parts.
- **Duralumin** is a mixture of 95% aluminium 4% copper and 1% magnesium. It has a low density so is light in weight but it is also strong and corrosion-resistant. It is used to make aircraft bodies.
- **Titanium** alloys contain mixtures such as titanium, aluminium and vanadium. The alloys are strong, light in weight and corrosion-resistant. They can withstand high temperatures and are used in the jet engines on aircraft.
- **Nichrome** is a mixture of nickel and chromium. It has a high melting point and is resistant to corrosion. It is used in the heating coils of electric fires.
- **Solder** is a mixture of 63% tin and 37% lead. It has a low melting point and is used to join wires in electrical circuits.
- **Steel** is mixed with other metals to make **alloy steels**.
- **Stainless steel** is an alloy of 73% iron, 1% carbon, 8% nickel and 18%

chromium. It has many of the properties of other steels but it does not rust.

- **Invar steel** is an alloy of nickel and steel which does not expand when the temperature changes. It is used for making very accurate measuring instruments.

(iv) Changing the properties of metals

The atoms in a metal are arranged in layers or sheets. When a force is applied to the metal the sheets of atoms slide over each other. This allows the metal to be pulled into a wire or rolled into a sheet.

The layers of atoms form crystals in the metal which are known as **grains**. There are imperfections or faults in some of the layers. These faults are called dislocations and they cause weaknesses in the crystal structure and consequently in the piece of metal.

When a force is applied to a metal the dislocations move. They are stopped at the edge of the crystal because the layers of atoms in the neighbouring crystals are arranged in different directions. The places where the edges of the crystals (grains) meet are called **grain boundaries**. A metal with large crystals (grains) has fewer grain boundaries than a metal with small crystals.

QUESTIONS

12 Draw a sample of a metal with small crystals by drawing small squares and rectangles joined together. Now make a second drawing of a metal with larger grains and note how it has fewer grain boundaries.

13 Which will be more malleable a metal with large grains or a metal with small grains? Explain your answer.

14 Metals with small grains are harder and stronger than metals with large grains. Why is this?

The strength and hardness of a metal can be altered by treating the metal in the following ways:

(a) Work hardening

A piece of metal is reshaped without heating it. The forces used to reshape the metal move the dislocations about in the grains until their positions block the movement of each other. This process in which heat is not used is also called **cold working**. It increases the brittleness of the metal.

(b) Quenching

The metal is heated then cooled very quickly by immersing or quenching it in cold water or oil. This rapid change in temperature causes a change in the crystal structure which in turn alters the properties of the metal.

The properties of the metals which have been introduced by cold working or quenching can be modified in the following processes to produce a metal with the exact properties required.

After a metal has been work hardened it can be treated by:

Annealing

In this process the metal is heated until its atoms can move and realign and new small crystals can form. This reduces the brittleness and strength of a work-hardened metal. The process of cold working and annealing may be repeated.

Forging

In this process the metal is heated so that it becomes soft and forms new crystals, then is reshaped as in cold working. This process is also known as **hot working**.

Tempering

After quenching a metal may be tempered by raising its temperature again but not as highly as before, then allowing it to cool in air.

QUESTION

15 Distinguish between the different kinds of metal-working techniques.

(c) Limestone → lime

Lime is used to remove slag in making steel. It is used to make mortar for binding bricks together. Other ingredients of mortar are cement (made by heating limestone and clay), sand and water. The lime helps the cement to form a stronger more weather-resistant bond between the bricks. Lime is used to purify water and to reduce the acidity in soils.

Lime is made by heating limestone in a kiln, as shown in Figure 16.6.

The symbol equation for the reaction which takes place is:

$$CaCO_3 \text{ (s)} \rightarrow CaO \text{ (s)} + CO_2 \text{ (g)}$$

$$CaCO_3(s) \rightarrow CaO(s) + CO_2(g)$$

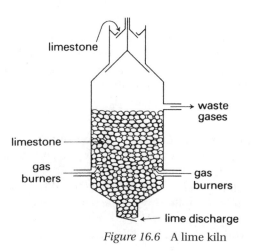

Figure 16.6   A lime kiln

(d) Clay → ceramics

There are many kinds of ceramics. They are non-metallic substances which have been heated strongly to form hard, brittle materials. Some ceramics are capable of withstanding very high temperatures and are used for lining furnaces used in the manufacture of other materials.

Clay is the most widely used ceramic material. When it is heated its particles become bonded together by a glassy substance which forms where the particles touch. Clay ceramics are used in pottery and electrical insulators.

(e) Sand → glass

Sand is the main raw material for making glass. The others are sodium carbonate (soda ash) and calcium carbonate (limestone). Sand is formed from silica. It combines with sodium and calcium to form sodium and calcium silicates. These are the main substances from which glass is made.

Glass bottles are made by placing a pellet of molten glass in a mould and blowing compressed air onto it to make it press against the sides of the mould. Plate glass for windows is made by floating a sheet of molten glass on a bath of molten tin. This process gives the sheet of glass a very smooth, shiny surface and a constant thickness along its length.

QUESTION

16 Assess the importance of limestone in the building industry.

(v) Oil products

(a) Petroleum

The uses of the oil products which were separated by fractional distillation are shown in Figure 16.4. When the products are separated it is found that there are not enough short chain molecules to meet our needs and that there are too many long chain molecues for our requirements. The long chain oil products are broken down into more useful smaller molecules by a process called **cracking**. In this process the long chain molecules are heated and passed over a catalyst which is a mixture of the oxides of aluminium and chromium at a temperature of about 500°C. Long chain molecules such as decane (10 carbon atoms) is broken down into ethene and octane; dodecane (12 carbon atoms) is broken down into decane and ethene.

QUESTION

17 Write symbol equations for the cracking of (a) decane, (b) dodecane. (Use the information about hydrocarbon structure on pp. 209 and 214 to help you.)

(vi) Polymers – structure and uses

The addition reaction which produces some polymers such as polyethene is described on p. 217.

Figure 16.7 Structural formula of part of a polyethene molecule

The structural formula of part of a polyethene molecule is shown in Figure 16.7.

This polymer is more commonly known as polythene. There are two kinds – **low-density polythene** and **high-density polythene**.

Low-density polythene has molecules arranged at random with relatively large spaces between them. It is used for plastic bags, sheets and films.

In high-density polythene parts of the molecules are laid side by side. It is tougher and more hard-wearing than low-density polythene and is used for bowls and pipes.

(a) Polypropene

This polymer has a CH_3 group in place of the hydrogen atom in the lower right position of Figure 16.7. It is tougher than polythene and is used to make fibres in some carpets and ropes and for making 'plastic' chairs.

(b) Polyphenylethene

This polymer has a C_6H_5 group in place of the right lower hydrogen atom in Figure 16.7. It is known as polystyrene and has a wide range of uses from packaging materials to 'plastic' toys.

(c) Polychloroethene

This polymer has a chlorine atom in place of the right lower hydrogen atom in Figure 16.7. It is known as PVC. It is used to make rainwear such as coats, hats and wellington boots. It is used to coat wires that carry electricity.

(d) Polytetrafluoroethene

This polymer has four fluorine atoms instead of the four hydrogen atoms in Figure 16.7. It is known as PTFE and is used in non-stick pans.

QUESTION

18 Draw the molecule structure of (a) polypropene, (b) polystyrene, (c) PVC, (d) PTFE.

(vii) Properties of polymers

The uses of polymers are related to their properties.

- Polythene is flexible and strong which makes it useful for bags and sheets. Its strength and waterproof properties make it suitable for kitchen bowls.
- Polypropene is hard and stiff. These properties make it useful for hard-wearing carpets.
- Polystyrene is strong which makes it useful for a packing material.
- PVC has waterproof and flexible properties which make it suitable for clothing and for covering the wire leads of electrical equipment.
- In addition to PTFE's non stick properties it is also heat-resistant, which makes it useful for lining pans.

(viii) Plastics and heat

A plastic material is a material usually made from a polymer which is liquid at some stage in its manufacture then sets to form a solid.

Plastics can be divided into two groups according to how they react to heat.

(a) Thermoplastics

The molecules in thermoplastics (which are also known as **thermosoftening plastics**) are not connected to each other. When the plastic is heated it softens then becomes hard again as it cools. Examples of thermoplastics are polythene, polystyrene and PVC. These materials can be remoulded by warming and cooling.

(b) Thermosetting plastics

The molecules in thermosetting plastics are firmly linked together so that they cannot move when they are heated. This lack of movement prevents the plastic from melting. Objects made from thermosetting plastic are manufactured from powders which are placed in a mould then pressed and heated. The polymers form and set in the shape made by the mould. Melamine used for making heat-proof surfaces and bakelite used for making pan handles are two examples of thermosetting plastics. These materials can never be remoulded.

QUESTION

19 Why would a hot pan melt a polythene surface but not one made of melamine?

(ix) General properties of plastics

Plastics are low density (light-weight) products. This makes them easy to handle. They can be made into a very wide range of shapes from simple sheets to containers with many compartments. Plastics are not corroded by chemicals and do not let water pass through them. They can be made to withstand strong pulling and pushing forces. Coloured substances are easily added to a plastic in the production process to make the finished product more attractive. By adding different atoms or groups of atoms to a polymer a plastic material can be designed to meet the needs of a particular task.

Nearly all plastics are not attacked by microorganisms and so do not decay. This can lead to pollution problems when the plastic products are thrown away.

Another property of plastics which can cause problems is their flammability. When they burn they may produce toxic gases which can poison people trying to escape from a burning building.

(x) Vegetable oils

(a) In food

Many vegetable oils are used in liquid form in cooking. They can also be solidified into margarine by the additon of hydrogen to their molecules. This is possible because the molecules in vegetable oils contain double bonds (a maximum of three per molecule) between some of their carbon atoms. The reaction is described on p. 216.

(b) Soap

If vegetable oils are heated with sodium hydroxide solution they form fatty acids and glycerol. When a strong solution of sodium chloride (brine) is added sodium stearate forms a precipitate which can be separated from the liquid by filtration.

Sodium stearate forms sodium and stearate ions. The stearate ions take part in the cleaning process. The sodium ions are spectator ions (see p. 81). The formula for the stearate ion is $C_{17}H_{35}CO_2^-$.

This is a very long molecule. The carboxylic (CO_2) end is called the water-loving head and the hydrocarbon $(C_{17}H_{35})$ end is called the grease-loving tail. The way these molecules are used to clean materials is described on p. 95.

(xi) Composite materials

A composite material is composed of two different materials. The combined properties of the materials give the composite special useful properties as the following examples show.

(a) Reinforced concrete

Concrete is made from mortar, sand and gravel. It can stand compressive (squashing) forces but is broken by small tension (bending) forces. When steel wires are added to the concrete before it sets the composite can withstand strong tension forces too. Reinforced concrete is use in making bridges and in the foundations of large buildings.

(b) Plastic reinforced glass

Sheets of glass are strong but very brittle and shatter into many pieces. If a car windscreen made of a sheet of glass is broken while the car is being driven the occupants of the car may be seriously injured and the car may go out of control and cause an accident. Car windscreens are made from two sheets of glass with a transparent sheet of plastic between them. If the sheets of glass are broken the plastic sheet holds the pieces in place.

(c) Glass reinforced plastic

Glass fibres are strong but they are brittle. If they are woven into a mat and coated in a plastic resin they form a strong light-weight material that does not break easily. It is water-resistant and not damaged by chemicals and is used to make some boat hulls and car bodies.

(d) Carbon fibre

Carbon fibres are very strong yet very light in weight. They are woven into a mat and set in a plastic resin to make a strong, stiff light-weight material which is water-resistant and not damaged by chemicals. It is used to make tennis rackets, fishing rods, and the bodies of racing cars and aircraft.

Alloys and ceramics are further example of composites.

(e) Natural composites

Wood and bone are two examples of natural composites. Wood is made from fibres of cellulose (a natural polymer). The fibres are bound together by a substance called **lignin**. Wood is a strong, flexible and light-weight material which has many uses.

Bone is made from fibres of a protein called collagen. They are held together by calcium phosphate to make a strong light-weight material. Bone makes the skeleton which is a supporting structure for the softer organs of the body. The bones in the limbs provide light-weight support of the muscles. If the bones were heavy large amounts of energy would be needed for movement.

QUESTION

20 Give an example of how a composite is an improvement on existing material to perform a particular task.

(xii) Making ammonia

Ammonia is used in the manufacture of fertilisers, nitric acid and plastic materials such as nylon and urea formaldehyde resins.

Ammonia is made by the Haber Process. This is named after **Fritz Haber** (1868–1934) who discovered a way of making ammonia from nitrogen and hydrogen. **Karl Bosch** (1874–1940) was the chemical engineer who scaled up Haber's laboratory process into an industrial process.

The raw materials for the Haber Process are air (which supplies the nitrogen) and naptha or natural gas and water (which provides the hydrogen).

(a) Preparation of nitrogen

Nitrogen is prepared by liquifying air by compression and cooling then fractionating it in a similar way to the fractionating of oil (see Figure 16.4). The nitrogen is collected in gaseous form from the top of the fractionating column.

(b) Preparation of hydrogen

If naptha is used it is heated with steam and the reaction takes place as shown in this symbol equation:

$$C_6H_{14} (g) + 12H_2O (g) \rightarrow 6CO_2 (g) + 19H_2 (g)$$

If natural gas is used the methane is heated with steam to produce hydrogen and carbon dioxide.

QUESTION

21 Write the symbol equation for the reaction between methane (CH_4) and steam (H_2O).

Carbon dioxide can be removed from the gas mixture in two ways. In one, the carbon dioxide can be removed by cooling and increasing the pressure on the gases so that it turns into a liquid while the hydrogen remains as a gas. In the second, the carbon dioxide can be removed by dissolving it in water or in an alkali.

(c) The Haber Process

Hydrogen and nitrogen are mixed together in a ratio of 3 volumes of hydrogen to 1 volume of nitrogen. This mixture is called **synthesis gas**. The pressure of the mixture is increased to about 200 atmospheres and the temperature is increased to about 400°C. The hot compressed gases are then passed over pellets of iron which act as a catalyst.

The symbol equation for the reaction is:

$$3H_2 (g) + N_2 (g) \rightleftharpoons 2NH_3 (g) \qquad \Delta H\text{-}92\,kJmol^-$$

As the reaction is reversible and all three gases are present in the mixture leaving the reaction chamber, the conditions are set to move the equilibrium position (see p. 200) to the right of the equation so that as much ammonia can be made as possible. Even by modifying the conditions for maximum production only 15% of the gas leaving the reaction chamber is ammonia. The gases are cooled so that the ammonia is liquified and can be separated. The remaining nitrogen and hydrogen then are mixed with more synthesis gas and sent through the reaction chamber again.

QUESTIONS

22 Why were iron pellets used for the catalyst and not an iron sheet (see p. 194)?

23 Is the reaction between nitrogen and hydrogen exothermic or endothermic (see pp. 201–2)?

(d) Ammonia and fertiliser

Most of the ammonia (81%) produced is used to make fertiliser. Two common fertilisers are **ammonium nitrate** and **ammonium sulphate**.

Ammonium nitrate is made by allowing an excess of dilute ammonia solution to mix with nitric acid. The excess ensures that as much ammonium nitrate can be made as possible. When the solution is heated any unreacted ammonia evaporates and the ammonium salt can be collected by crystallisation. The symbol equation for the reaction is:

$$NH_3 (aq) + HNO_3 (aq) \rightarrow NH_4NO_3 (aq)$$

QUESTION

24 Construct an equation for the reaction between ammonia solution and dilute sulphuric acid.

(e) Nitric acid

Nitric acid is used to make ammonium nitrate fertiliser, synthetic fibres such as nylon, explosives such as TNT and dyes.

The raw materials are air and ammonia and water.

The manufacture of nitric acid takes place in three stages:

Stage 1: Production of nitrogen monoxide

Air and ammonia are mixed. The quantities are 90% air and 10% ammonia. The mixture is heated to about 230°C and passed over a catalyst made of platinum with a small amount of rhodium. The reaction between oxygen in the air and ammonia is exothermic and raises the temperature of the mixture to 800°C. The symbol equation for this reaction is:

$$4NH_3 (g) + 5O_2 (g) \rightarrow 4NO (g) + 6H_2O (l)$$

Stage 2: Production of nitrogen dioxide

Nitrogen monoxide is a colourless gas. When it reacts with oxygen in the air a brown gas nitrogen dioxide is produced. The symbol equation for this reaction is:

$$4NO (g) + 2O_2 (g) \rightarrow 4NO_2 (g)$$

Stage 3: Production of nitric acid

The nitrogen dioxide reacts with oxygen and water to form nitric acid, as the following symbol equation shows:

$$4NO_2 (g) + 2H_2O (l) \rightarrow 4HNO_3 (aq)$$

A very small amount of nitrogen dioxide is left over from the reaction and is released into the atmosphere.

(f) Properties of nitric acid

Nitric acid is a mineral acid. The properties of mineral acids are described in Chapter 11.

(xiii) Making sulphuric acid

Sulphuric acid is used in batteries in cars and trucks, in fertilisers (see Question 24, p. 239) in paints, detergents, for making fibres and plastics and in the cleaning of newly made metals.

Sulphur dioxide, air and water are the raw materials for the manufacture of sulphuric acid.

(a) The production of sulphuric acid

Sulphur dioxide can be produced:

- By burning sulphur which is obtained from underground deposits.
- By burning sulphur which is obtained in the refining of fossil fuels such as natural gas and oil.

The chemical equation for this reaction is:

$$S\,(s) + O_2\,(g) \rightarrow SO_2\,(g)$$

The reaction is exothermic and the heat released is used to melt the solid sulphur.

- From the first stage in the extraction of zinc from zinc sulphide and the extraction of lead from lead sulphide. The following symbol equation is for the first stage in the extraction of zinc:

$$2ZnS\,(s) + 3O_2\,(g) \rightarrow 2ZnO\,(s) + 2SO_2\,(g)$$

QUESTION

25 Construct a symbol equation for the first stage in the extraction of lead (Pb) from lead sulphide (PbS).

(b) The contact process

Air and sulphur dioxide are heated to 450°C and passed over a vanadium (V) oxide catalyst. The reaction between oxygen in the air and the sulphur dioxide is shown in the following symbol equation:

$$2SO_2\,(g) + O_2\,(g) \rightleftharpoons 2SO_3\,(g)$$

The reaction is exothermic and reversible.

The temperature and the pressure of the reactants is controlled to move the equilibrium position to the right so that a very large amount of sulphur trioxide is produced.

(c) Formation of sulphuric acid

As the reaction between sulphur trioxide and water is very violent the gas is dissolved in concentrated sulphuric acid. The product of this reaction is oleum, as the following equations show:

sulphuric acid + sulphur trioxide $\rightarrow$ oleum

$$H_2SO_4\,(aq) + SO_3\,(g) \rightarrow H_2S_2O_7\,(l)$$

The oleum is then added to water to make sulphuric acid, as the following symbol equation shows:

$$H_2S_2O_7 \text{ (l)} + H_2O \text{ (l)} \rightarrow 2H_2SO_4 \text{ (l)}$$

The sulphuric acid has a concentration of 98%.

(d) Acid production and pollution

In recent years the pollution of the air due to the release of sulphur dioxide not used in the reaction has been greatly reduced by making the manufacturing process more efficient.

(e) Properties of sulphuric acid

Sulphuric acid is a mineral acid. The properties of mineral acids are described in Chapter 11.

The properties described here are those of concentrated sulphuric acid only.

Sulphuric acid and water

When sulphuric acid and water are mixed an exothermic reaction takes place. Water is less dense than sulphuric acid and if it is added to the acid it forms a layer on top. The heat from the reaction makes the water layer boil which causes drops of the acid to rise into the air. To avoid this danger the acid should be added to the water. The mixture is heated as the acid sinks through the water but it does not boil.

Sulphuric acid and the water of crystallisation

Sulphuric acid can be used to remove the water of crystallisation from copper (II) sulphate crystals. The symbol equation for this reaction is:

$$CUSO_45H_2O \text{ (s)} + H_2SO_4 \text{ (l)} \rightarrow CuSO_4 \text{ (s)} + H_2SO_4 \text{ (aq)}$$

QUESTION

26 What colour change takes place in this reaction (see also p. 156)?

(f) Drying gases

Sulphuric acid can be used as a drying agent to remove water from moist gases with which it does not react such as nitrogen, oxygen and chlorine. It cannot be used to dry ammonia.

(g) Turning sugar to carbon

Sulphuric acid can remove water from a molecule of sucrose (a sugar) leaving carbon behind. The water mixes with the acid to make it more dilute. The symbol equation for this reaction is:

$$C_{12}H_{22}O_{11} \text{ (s)} + H_2SO_4 \text{ (l)} \rightarrow 12C \text{ (s)} + H_2SO_4 \text{ (aq)} + 11H_2O \text{ (l)}$$

16.3 Summary

- Iron is extracted from its ore in a blast furnace (see p. 223).
- Electrolysis is used in the extraction of aluminium (see p. 225).
- Electrolysis is used in the purification of copper (see p. 226).
- A wide range of products are produced by the fractional distillation of oil (see p. 228).
- The properties of iron and steel depend on the amount of carbon they contain (see p. 228).
- Alloys are mixtures of different metals (see p. 230).
- Metal-working techniques change the properties of metals (see p. 231).
- Lime is made by heating limestone and is used for making mortar and reducing acidity in soil (see p. 232).
- Ceramics are made by heating clay (see p. 233).
- Glass is made by heating sand, limestone and soda ash (see p. 233).
- Small hydrocarbon molecules can be made by cracking large hydrocarbon molecules (see p. 233).
- Polymers can be made which have specific useful properties (see p. 234).
- There are two kinds of plastic – thermoplastics and thermosetting plastics (see p. 235).
- Soap can be made from vegetable oils and sodium hydroxide solution (see p. 236).
- A composite material is made from two different materials and has properties which differ from both of them (see p. 236).
- Ammonia is manufactured by the Haber Process (see p. 238).
- Nitric acid is manufactured in a three-stage process (see p. 239).
- Sulphuric acid is manufactured by the contact process (see p. 240).
- Sulphuric acid is a dehydrating agent (see p. 241).

■ ⋈ 17 Chemistry and the environment

Objectives

When you have completed this chapter you should be able to:
- Explain how **raw materials** are collected
- Understand the important factors in selecting a site for a **chemical plant**
- Understand the need for **efficiency** in the production of materials
- Explain the **advantages** of using fertiliser and pesticides
- Describe the **consequences** of the misuse of fertilisers and pesticides
- Understand how **acid rain** is produced and the damage that it can cause
- Describe how water can be **polluted**
- Describe how **air pollution** is caused and the steps that can be taken to reduce it
- Understand how the Earth's resources must be **managed**.

We rely on large-scale chemical processes to provide us with the materials we need to maintain our current lifestyle. While the chemical reactions themselves do not produce environmental damage, the way we collect the raw materials and deal with the wastes and the products when we no longer need them has led to many forms of harmful pollution.

17.1 Raw materials

(i) Solids

Metal ores such as haematite, rocks such as limestone, minerals such as diamond and coal (which is a fuel and a source of chemicals) are removed from the ground by mining processes. Many raw materials are removed from the Earth's crust by digging tunnels into it. This form of mining is expensive and there is a danger of a collapse in the mine. However, it causes a relatively small amount of damage to the surrounding habitats. In open-cast mining which is cheaper and safer for miners than making tunnels the overlying surface is removed to extract the raw material.

When the open-cast mining operation is complete the surface can be replaced. In temperate regions such as Europe the natural woodland habitat may be restored but in tropical rainforest areas the thin soil is washed away and prevents the previous habitat being reestablished.

(ii) Liquids

Oil is a raw material for many products. It is extracted from underground through pipes. The major risk to the environment is in the transportation of oil by sea. It is transported in huge ships called tankers. If a tanker runs aground or has a collision with another ship it may spring a leak. The escaping oil causes damage to water life at the sea's surface and on the shore. In the past oil tankers were cleaned out by releasing oil into the sea. Today any oil which remains from one consignment is kept on board and transported with the next consignment.

Water takes part in many chemical reactions but it is also used to transport heat energy. It may in the form of steam carry heat energy to a reaction. Water may also be used to remove heat (see also p. 138).

(iii) Gases

The air is a raw material. It is liquified by increasing its pressure and lowering its temperature, then the components of the air are separated by fractional distillation, as described for oil on p. 227. The noble gases are separated and put to a range of uses. Oxygen may be kept in liquid form and used in the combustion of rocket fuel on space craft. Nitrogen is combined with hydrogen to form ammonia in the Haber Process.

QUESTION

1 Assess the environmental damage caused in collecting raw materials.

17.2 Finding a site for a chemical plant (Figure 17.1)

A chemical plant costs a huge amount of money to build and to run. It must be set up in a place where its products can be made cheaply so that large amounts can be sold to pay off the debt of construction, the wages of the workforce and to make a profit for the company running the plant.

The product can be made cheaply if the raw materials can be delivered to the plant easily and the products can be transported to the customers easily. Towns are needed close by to provide a workforce for the plant.

Early chemical and industrial plants were set up at the site of the raw material or a fuel and towns were set up around them. For example, in Yorkshire (UK) towns were set up so that most of the populations could work in the iron and steel industry that had developed close to nearby sources of iron and coal. The local coal was made into coke for the blast furnaces.

Today the selection of a site for a chemical plant depends upon a range of factors.

Figure 17.1 A chemical plant

(i) Transport of raw materials

Many raw materials for processing in a country come from other parts of the world. They are transported in large quantities in ships for cheapness. This means that many processing plants need to be either near the coast or on the side of a wide, deep river in which the ships can moor to offload their cargo.

(ii) Transport of products

The ease with which a rail or road transport system can distribute the products to the rest of the country is an important factor in site selection.

(iii) Nearby towns

The presence of nearby towns is important for providing a workforce. The ease with which the workforce can travel between the towns and the plant with a minimum of environmental damage is also a consideration.

In most places there is a prevailing wind. It is the wind which most frequently blows in a certain direction. In the past many chemical plants released their waste products directly into the air with the result that people living down wind of the plant suffered from air pollution. Today there are stricter controls on the release of wastes but if an accident occurred there is still a risk of pollution. To minimise this risk the site should be selected so that there are no towns down wind of the plant.

(iv) Electricity

If large amounts of electricity are required, as in the extraction of aluminium, a power station is needed close by to provide cheap electricity. Hydro-electric schemes involving water storage in reservoirs and water turbines to turn the generators provide cheap electricity.

(v) Existing land use

The land for the site should not be valuable agricultural land or the habitat of rare organisms. It should also not be on important migration routes where migrants gather, such as the winter feeding grounds of European geese.

QUESTION

> 2 Look at a map of your area. Where would be the best place to site a chemical plant, assuming that all the raw materials were going to have to be brought from other places?

17.3 Efficiency

Once a plant has been set up and is running, research continues to find ways to make the processes and operations at the plant more efficient. For example, ways may be found to use smaller amounts of expensive catalysts or to use a cheaper fuel. Many of the reactions are exothermic and the heat produced may be used in other parts of the plant or sold to neighbouring industrial plants. Where a product of one industrial plant is a raw material for a second industrial plant the two plants are set up close together to minimise transport costs. An example of two plants put close together for this purpose are the plants producing ammonia and nitric acid (see pp. 237 and 239).

The waste products of a process may also have a use in making other materials – for example, the metals in the anode sludge produced in the purification of copper (see p. 226).

By increasing efficiency and selling waste products for use in other industries a chemical plant can keep down the cost of its products and sell at a competitive price to other chemical plants around the world. This allows the product to be used widely, it keeps the workforce employed and makes profits for the company which can be used to finance further research and development.

17.4 Useful chemicals in the environment

(i) Fertilisers

Protein is the major structural substance in cells of plants and animals. It also forms enzymes (see p. 192) which catalyse chemical reactions of life processes. Nitrogen is an essential component of protein. It enters the food chain as **nitrate ions** which pass from the soil water into plant roots.

The plants use the carbohydrate they make in photosynthesis to make other materials such as protein. This passes along the food chain to animals including humans (Figure 17.2).

The supply of nitrates in the soil is quite low in nature but adequate for the relatively small human population of the past. Extra nitrogen was added by the use of manure. With the rapid increase in the human population the demand for more food, and consequently more nitrates in the soil, have increased to supply the extra protein needs. This demand could not be met by the use of manure but has been met by the development of fertilisers such as ammonium nitrate and ammonium sulphate. There are many other elements required by plants but phosphorus and potassium are needed in particularly large amounts. These are also provided by fertilisers. The fertiliser ammonium phosphate provides both nitrogen and phosphorus at the same time and potassium is provided by potassium sulphate or potassium chloride.

The use of fertilisers allows crops to grow to their full potential and not be stunted by lack of nutrients. The amount of food that a crop produces is called its yield. By developing new varieties of crop plants which produce large and more numerous grains, for example, and using fertilisers, crop yields can be greatly increased.

QUESTION

3 We eat cereals such as wheat and maize. What would be the effect on such crops if they were grown year after year in the same ground if manure or fertiliser was not used?

Figure 17.2 Harvesting a cereal crop

As food production has become more efficient, less land is needed for agriculture.

(ii) Pesticides

Some species of insects feed on crop plants. They can form large infestations which reduce the crop's yield. Chemicals (insecticides) have been developed which can kill off insect pests and protect crop yields.

Some species of fungi infect crop plants and cause them to decay. This reduces the crop yield. Chemicals (fungicides) have been developed to kill off these fungi and protect crop yields.

(iii) Lime

Lime has many uses but it is used in the environment to reduce the acidity of soil, which in turn leads to greater crop production.

17.5 Pollution

Despite improvements in the manufacturing of many materials pollution remains a global problem which if left unchecked may threaten life on Earth.

(i) Pollution on land

(a) Pesticides

An insecticide called DDT enters the food chain and becomes concentrated in the tissues of animals at the food chain top. This can have fatal results. DDT is not used in developed countries today, it has been replaced by other insecticides. Careless use of pesticides can affect food chains in the natural habitats around crop fields.

QUESTION

4 Issues concerning pesticides are in the news. Scan newspapers daily for a few weeks and collect any articles relating to the use of pesticides. How do these articles affect your views on the use of pesticides?

(b) Plastics

As plastics do not decay space is needed to store them when the objects they have been used for are no longer needed. This increases the demand for landfill sites for rubbish dumps and threatens natural habitats.

(c) Acid rain

When acid rain reaches the soil it removes minerals that plants need and produces aluminium sulphate which damages tree roots. The trees are weakened by the lack of nutrients and the root damage. This makes them more susceptible to attack by pathogenic microorganisms and they develop diseases which eventually kill them.

(ii) Pollution in water

(a) Pesticides

DDT washed from crop fields into rivers and the sea causes damage to aquatic food chains as it does to food chains on land.

(b) Over-use of fertilisers

The amount of fertiliser for use on a crop must be carefully calculated. If too much is used it is wasted as it washed out of the soil by rain and enters streams and rivers. Here it is taken up by water plants where it produces a great increase in growth. When the plants die they are decomposed by bacteria which take oxygen from the water. As there is a huge amount of plant material vast numbers of bacteria develop to break it down. The bacteria take so much oxygen from the water that other water organisms 'suffocate' and die. Damage to freshwater habitats by fertilisers is called **eutrophication**.

(c) Hot water

The waste heat from industrial processes and power stations is released in hot water. This raises the temperature of the river water into which it flows. The amount of oxygen dissolved in the water decreases as the temperature rises causing water organisms to die.

(d) Industrial wastes

Some industrial wastes cannot be treated at a sewage works because they damage the populations of microorganisms that are used there. These wastes may be released directly into a river where they are diluted. However, the chemicals are released in such large amounts at regular intervals that their concentrations in the aquatic habitats rises. Compounds such as mercury and organic compounds called PCBs can enter food chains and damage populations of organisms in them including human populations.

(e) The effect of acid rain

The pH of freshwater in regions where acid rain falls is reduced. This causes aquatic organisms to die.

(iii) Pollution of the air

(a) The formation of acid rain

Rainwater is made slightly acidic by the carbon dioxide which dissolves in it (see p. 75). The acidity is greatly increased by oxides of two other non-metals. Many power stations use fossil fuels to produce heat for the steam to spin turbines and make generators produce electricity. The fossil fuels contain small amounts of sulphur which is converted into sulphur dioxide as the fuel burns. When this gas reaches the atmosphere it dissolves in water droplets in clouds to form sulphurous acid, as the symbol equation shows:

$$SO_2 \text{ (g)} + H_2O \text{ (l)} \rightleftharpoons H_2SO_3 \text{ (aq)}$$

The high temperature in the furnace of a power station causes a reaction between nitrogen and oxygen in which nitrogen oxides are formed. These too form acids, as the example in the following equation show:

nitrogen dioxide + water → ions of nitric and nitrous oxides

$$2NO_2 \text{ (g)} + H_2O \text{ (l)} \rightarrow 2H^+ \text{ (aq)} + NO^{3-} \text{ (aq)} + NO^{2-} \text{ (aq)}$$

When acid rain falls to the ground it causes damage on land (see p. 248) and in fresh water (see p. 249).

(b) Cleaning industrial gases

Increasing efficiency

The fluid bed combustion method is a more efficient way of burning coal and produces smaller amounts of nitrogen oxides.

The amount of sulphur dioxide produced by coal-fired power stations can be reduced by simply using coal which has a smaller sulphur content.

More expensive ways of cleaning the gases involves a process called **scrubbing**.

Ammonia scrubbing

The sulphur dioxide and nitrogen oxide gases are made to react with air and water to form sulphuric acid and nitric acid. These acids are then neutralised by ammonia gas to form ammonium sulphate and ammonium nitrate. These products can then be used as fertilisers.

QUESTION

5 Produce word and symbol equations for the reactions between ammonia and (a) sulphuric acid (H_2SO_4), (b) nitric acid (HNO_3) in the ammonia scrubbing process.

Lime scrubbing

Sulphur dioxide is passed through lime and calcium sulphate is produced. This is a compound that is used to make plaster for walls. If the product of the scrubbing process can be made to the purity needed for plaster work, the expense of fitting the flue gas desulphurisation (FGD) unit in which the lime scrubbing takes place could be covered by the sale of the plaster.

(c) Pollution from exhaust gases

The hydrocarbons in petrol take place in a combustion reaction in car and truck engines which produce exhaust gases. These are carbon monoxide and nitrogen oxide. Carbon monoxide is a deadly gas in concentrations as low as 0.1%. Nitrogen oxide is a gas which dissolves in water in the atmosphere to form acid rain.

These gases are removed from the car exhaust by a catalytic converter. This has a honeycomb structure coated in platinum and rhodium. The following equation shows how carbon monoxide and nitrogen oxide react in the presence of the hot catalyst:

$$2CO \ (g) + 2NO \ (g) \rightarrow 2CO_2 \ (g) + N_2 \ (g)$$

(d) The greenhouse effect

When heat from the Sun reaches the Earth some is absorbed and some is reflected. The carbon dioxide in the atmosphere prevents some of this heat escaping in the same way that glass prevents heat reflected from inside a greenhouse passing back into the outside air. The combustion reactions in power stations and vehicle engines and, to a much smaller extent, the decomposition of limestone to produce lime increases the amount of carbon dioxide in the air. Rainforests are being cleared by burning. This also produces large amounts of carbon dioxide but also reduces the global plant life which can take in carbon dioxide through photosynthesis. It is believed that the increase in carbon dioxide concentration will cause more heat energy to be held in the atmosphere and cause global warming. This would produce a melting of the polar ice, flooding of low lands near the coasts and changes in climates in all regions.

The amount of carbon dioxide entering the air could be reduced by using fewer power stations burning fossil fuels and by using more power stations which use nuclear fuel, more hydro-electric power stations and more wind turbines in 'wind farms'. The use of electric vehicles instead of those using petrol or diesel would further reduce the amount of carbon dioxide in the air.

QUESTION

6 Issues concerning global warming are in the news. Scan newspapers daily for a few weeks and collect any articles relating to global warming, nuclear fuels and alternative energy sources. How do these articles affect you views on the causes of global warming and remedies to reduce it?

(e) The holes in the ozone layer

Between 25 and 30 kilometres above the surface of the planet is the ozone layer. It prevents harmful ultra violet radiation from the Sun reaching the ground.

Chlorfluorocarbons (CFCs) have been widely used in aerosol sprays and in refrigerators. They are inert gases and their escape into the atmosphere was thought to be harmless. However, when the gases reached the highly reactive ozone gas in the presence of ultra violet radiation chemical reactions took place which converted the ozone into oxygen. This has resulted in a thinning of the ozone layer and the development of holes in the layer over the north and south poles. Today CFCs are no longer used in developed countries. More stable compounds are now used.

17.6 Using the Earth's resources

There are two kinds of resources – **renewable** resources such as wood and **non-renewable** resources such as coal or a metal ore.

Renewable resources can be managed by balancing the regrowth of the material with its consumption for industry.

Non-renewable resources must be conserved to last as long as possible.

The world reserves of non-renewable resources are estimated by geological surveys made on the ground and by pictures and measurements made from satellites. From this information tables are produced which predict how long a non-renewable resource will last if it is used up at the present rate. Many of these predictions show that most of the non-renewable resources will be used up in less than 200 years and some could be used up in your life time.

From this research it is clear that industry must develop ways of conserving both raw materials and fossil fuels. One way is to improve efficiency (see p. 246) and another way is to develop re-cycling programmes. These already exist for iron, aluminium and glass and ways are being developed to recycle plastics (Figure 17.3).

When a material is recycled less energy is needed because energy is not needed to extract it (as metals from ores) or make it (as in the formation of glass).

A third way of conserving materials is to develop new materials from renewable resources to replace those currently made from non-renewable resources.

Figure 17.3   Recycling metals from scrapped cars

7 Issues concerning recycling, new materials and environmental damage are in the news. Scan newspapers daily for a few weeks and collect any articles relating to these topics. How can you use these articles to develop a positive attitude to the future of the planet?

17.7 Summary

- Solid raw materials are collected by mining (see p. 243).
- Oil and water are important raw materials (see p. 244).
- Air is a source of nitrogen and noble gases (see p. 244).
- Several factors are important in the selection of a site for a chemical plant (see p. 244).
- The cost of a product can be reduced by efficiency and selling the waste products (see p. 246).
- Fertilisers and pesticides help provide the crop yields we need (see p. 246).
- Fertilisers cause eutrophication (see p. 249).
- Pesticides can cause environmental damage (see p. 248).
- Water is polluted by pesticides, hot water, fertilisers and industrial wastes (see p. 249).
- Acid rain can be reduced by treating the smoke from a power station (see p. 250).
- A catalytic converter reduces the pollution in vehicle exhaust gases (see p. 250).
- The greenhouse effect causes the warming of the planet (see p. 251).
- The ozone layer is damaged by CFCs (see p. 251).
- There are several ways to conserve non-renewable resources for the future (see p. 252).

◼ ⌄ Answers to questions

Answers are given only to the numerical questions and questions on the construction of word and symbol equations

Chapter 1
7   $100\,cm^3$
8   3 atmospheres
9   (a) $293\,K$, (b) $308\,K$, (c) $523\,K$, (d) $263\,K$, (e) $216\,K$
10   (a) $27°C$, (b) $77°C$, (c) $227°C$, (d) $-73°C$, (e) $-263°C$
11   $2.34\,l$
12   $409.5\,K$
13   $1.94\,l$
14   $6.6\,l$
15   $18\,l$
16   $2.625\,l$
17   $120\,K$
18   $336\,K$

Chapter 4
Answer to Question 4:

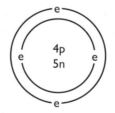

Chapter 6
1   (a) magnesium + oxygen → magnesium oxide
    (b) sodium hydrogen carbonate → sodium carbonate + water + carbon dioxide
    (c) barium chloride + sodium sulphate → barium sulphate + sodium chloride
2   carbonic acid → carbon dioxide + water
3   (a) $AgNO_3$, (b) $Ca(OH)_2$, (c) $Zn_3(PO_4)_2$, (d) $Al(OH)_3$, (e) $AgCl$, (f) $CaSO_4$, (g) $Fe(OH)_2$
4   (a) SiO_2, (b) PCl_3
5   (a) calcium and chlorine (two chlorine atoms to one calcium atom)
    (b) hydrogen and chlorine (one atom of hydrogen to one atom of chlorine)
    (c) nitrogen and hydrogen (one atom of nitrogen to three atoms of hydrogen)
    (d) sulphur and oxygen (one atom of sulphur to two atoms of oxygen)

(e) carbon and hydrogen (one atom of carbon to four atoms of hydrogen)

(f) carbon and hydrogen (one atom of carbon to three atoms of hydrogen)

6 (1) sodium + water → sodium hydroxide + hydrogen

(2) calcium carbonate → calcium oxide + carbon dioxide

(3) sodium hydroxide + nitric acid → sodium nitrate + water

11 (a) $2Cu$ (s) + O_2 (g) → $2CuO$

(b) H_2 (g) + Cl_2 (g) → $2HCl$ (g)

(c) $2NaHCO_3$ (s) → Na_2CO_3 (s) + H_2O (l) + CO_2 (g)

(d) Ag_2O (s) + H_2O_2 (aq) → $2Ag$ (s) + H_2O (l) + O_2(g)

(e) $2Al$ (s) + $3Cl_2$ (g) → $2AlCl_3$ (s)

12 (a) $2K$ (s) + $2H_2O$ (l) → $2KOH$ (aq) + H_2 (g)

(b) $MgCO_3$ (s) + $2HCl$ (aq) → $MgCl_2$ (s) + H_2O (l) + CO_2 (g)

(c) Fe_2O_3 (s) + $3CO$ (g) → $2Fe$ (s) + $3CO_2$ (g)

13 (a) the sodium and sulphate ions

(b) Fe^{2+} (aq) + $2OH^-$ (aq) → $Fe(OH)_2$

14 (a) $2Cl^- - 2e^- → Cl_2$

(b) $Cu^{2+} + 2e^- → Cu$

(c) $2I^- - 2e^- → I_2$

Chapter 7

22 729°C

Chapter 8

1 $2Zn$ (s) + O_2 (g) → $2ZnO$ (s)

2 potassium + water → potassium hydroxide + hydrogen

$2K$ (s) + $2H_2O$ (l) → $2KOH$ (aq) + H_2 (g)

3 (a) sodium + sulphuric acid → sodium sulphate + hydrogen

$2Na$ (s) + H_2SO_4 (aq) → Na_2SO_4 (aq) + H_2 (g)

(b) calcium + sulphuric acid → calcium sulphate + hydrogen

Ca (s) + H_2SO_4 (aq) → $CaSO_4$ (aq) + H_2 (g)

4 (a) silver oxide → silver + oxygen

$2Ag_2O$ (s) → $4Ag$ (s) + O_2 (g)

(b) calcium hydroxide → calcium oxide + water

$Ca(OH)_2$ (s) → CaO (s) + H_2O (l)

(c) copper carbonate → copper oxide + carbon dioxide

$CuCO_3$ (s) → CuO (s) + CO_2 (g)

7 The sulphate ions are the spectator ions

Zn (s) + Cu^{2+} (aq) → Zn^{2+} (aq) + Cu (s)

14 (a) $2CuO$ (s) + C (s) → $2Cu$ (s) + CO_2 (g)

(b) PbO (s) + C (s) → Pb (s) + CO_2 (g)

Chapter 9

19 Cl_2 (g) + Na_2S (aq) → S (s) + $2NaBr$ (aq)

Chapter 10

16 $Mg(HCO_3)_2$ (aq) → $MgCO_3$ (s) + CO_2 (g) + H_2O (l)

19 $CaSO_4$ (aq) + $Na_2(CO_3)_2$ (aq) → $CaCO_3$ (s) + Na_2SO_4 (aq)

Chapter 11

1 (a) hydrochloric acid + iron → iron chloride + hydrogen

     $2HCl\ (aq) + Fe\ (s) \rightarrow FeCl_2\ (aq) + H_2\ (g)$

   (b) hydrochloric acid + calcium carbonate → calcium chloride + carbon dioxide

     $2HCl\ (aq) + CaCO_3\ (s) \rightarrow CaCl_2\ (aq) + CO_2\ (g)$

3 $H^+\ NO_3^-, H_3O^+\ NO_3^-$

8 (a) $Mg\ (s) + 2HCl\ (aq) \rightarrow MgCl_2\ (aq) + H_2\ (g)$

   (b) $Zn\ (s) + H_2SO_4\ (aq) \rightarrow ZnSO_4\ (aq) + H_2\ (g)$

10 (a) $MgO\ (s) + 2HCl\ (aq) \rightarrow MgCl_2\ (aq) + H_2O\ (l)$

   (b) $ZnO\ (s) + H_2SO_4\ (aq) \rightarrow ZnSO_4\ (aq) + H_2O\ (l)$

12 (a) $MgCO_3\ (s) + 2HCl\ (aq) \rightarrow MgCl_2\ (aq) + H_2O\ (l) + CO_2\ (g)$

   (b) $Na_2CO_3\ (s) + 2HCl\ (aq) \rightarrow 2NaCl\ (aq) + H_2O\ (l) + CO_2\ (g)$

13 (a) $2KOH\ (aq) + H_2SO_4\ (aq) \rightarrow K_2SO_4\ (aq) + 2H_2O\ (l)$

   (b) $NaOH\ (aq) + HNO_3\ (aq) \rightarrow NaNO_3\ (aq) + H_2O\ (l)$

Chapter 12

1 (a) 17, (b) 40, (c) 58.5, (d) 84, (e) 98

2 (a) 12%, (b) 48%, (c) 82.4%, (d) 39.3%, (e) 32.7%

3 (a) 4 mol, (b) 0.15 mol

4 5 mol.

5 33.3 mol

6 6 mol

7 (a) 18 g, (b) 378 g, (c) 20 g

8 (a) CO_2, (b) FeS, (c) $CaCl_2$

9 (a) SO_3, (b) NO_2, (c) $FeCl_3$

10 C_4H_8

11 28 g

12 44.8 g

13 4 mol/l

14 2 mol/l

15 (a) 1.25 mols, (b) 27 mols, (c) 10 mols

16 (a) 480 g, (b) 1200 g

17 70 g

18 4.8 M

19 1.67 M

20 (a) $4.8\ dm^3$, (b) $72\ dm^3$

21 (a) $12\ dm^3$, (b) $48\ dm^3$

22 $48\ dm^3$

Chapter 13

8 $NaCl\ (aq) + H_2O\ (l) \rightarrow NaOH\ (aq) + H_2\ (g) + Cl_2\ (g)$

9 (a) at anode $2Br^-\ (aq) - 2e^- \rightarrow Br_2\ (g)$, at cathode $Zn^{2+}\ (aq) + 2e^- \rightarrow Zn\ (s)$

   (b) at anode $OH^- - e^- \rightarrow OH\ (aq)$, $4OH\ (aq) \rightarrow 2H_2O\ (l) + O_2\ (g)$

     at cathode $2H^+ + 2e^- \rightarrow H_2\ (g)$

11 (a) 30 coulombs, (b) 60 coulombs, (c) 1800 coulombs

12 (a) 0.158 g, (b) 0.034 g

13 0.066 g
14 33.8 cm^3
15 3
16 (b)
17 0.5
18 81 g

Chapter 14

4 (a) Mg (s) + 2HCl (aq) $\rightarrow$ MgCl$_2$ (aq) + H$_2$ (g)
5 Increase in surface area of 24 cm^2 from 24 cm^2 to 48 cm^2
6 Increase in surface area to 96 cm^2
16 CuSO$_4$·5H$_2$O (s) $\rightleftharpoons$ CuSO$_4$ (s) + H$_2$O (l)
17 NH$_4$Cl $\rightleftharpoons$ NH$_3$ (g) + HCl (g)

Chapter 15

1 (a) C$_8$H$_{18}$, (b) C$_{18}$H$_{38}$
3 C$_6$H$_{14}$
12 (b) n(C$_6$H$_5$HC=CH$_2$) (l) $\rightarrow$ –(C$_6$H$_5$CH–CH$_2$)–n (s)
18 C$_2$H$_5$OH (l) + 3O$_2$ (g) $\rightarrow$ 2CO$_2$ (g) + 3H$_2$O (l)

Chapter 16

1 carbon + oxygen $\rightarrow$ carbon monoxide
carbon monoxide + iron oxide $\rightarrow$ iron + carbon dioxide
calcium carbonate $\rightarrow$ calcium oxide + carbon dioxide
calcium oxide + silicon dioxide $\rightarrow$ calcium silicate
7 TiCl$_4$ (l) + 4Na (l) $\rightarrow$ Ti (s) + 4NaCl (s)
17 (a) C$_{10}$H$_{22}$ (l) $\rightarrow$ C$_2$H$_4$ (g) + C$_8$H$_{18}$ (l)
(b) C$_{12}$H$_{26}$ (l) $\rightarrow$ C$_2$H$_4$ (g) + C$_{10}$H$_{22}$ (l)
21 CH$_4$ (g) + 2H$_2$O (g) $\rightarrow$ CO$_2$ (g) + 4H$_2$ (g)
24 NH$_3$ (aq) + H$_2$SO$_4$ (aq) $\rightarrow$ (NH$_4$)$_2$SO$_4$ (aq)
25 2PbS (s) + 3O$_2$ (g) $\rightarrow$ 2PbO (s) + 2SO$_2$ (g)

Chapter 17

5 (a) 2NH$_3$ (aq) + H$_2$SO$_4$ (aq) $\rightarrow$ (NH$_4$)$_2$SO$_4$ (aq)
(b) NH$_3$ (g) + HNO$_3$ (aq) $\rightarrow$ NH$_4$NO$_3$ (aq)

■ⱴ Index

A

acid
  dilute, reaction with metals   110
  rain   248, 249–51
  simple test for   150
actinide   86, 87
activation energy   203
addition reaction   74
alkali   153
alkaline earth metal   97–8
alkaline solution   92
alloy   223, 230
aluminium   112, 225–6, 230
ammonia   19, 125, 141, 237–9, 250
amphoteric oxide   129
anion   141, 143, 174
anode   174
anodising   183
antifreeze   217
aqua regia   108
argon   91
atmosphere   19
atom   50
atomic number   53–4, 85
Avogadro Law   170
Avogadro number   161

B

back reaction   75, 200
balanced equation   79
basalt   23, 29, 33
basic metal oxide   129
batholith   23
bauxite   225
'Big Bang'   1, 18
blast furnace   224, 244
boiling point   3, 47, 92, 97, 99
Boyle's Law   7
bromine   99, 100, 147, 175, 216
Brownian motion   11, 13

C

calcium   97–8, 109, 145, 146
carbon   18, 115, 119–23, 223
carbon cycle   122
carbon dioxide   19, 75, 120–2, 144
carbon monoxide   19, 120

catalyst   105, 190–2, 196, 216, 238, 239, 240
cathode   174, 175
cation   141, 175
centrifuge   40
CFCs   251
chalk   24
Charles' Law   8
chemical compound   59
chlorides   108
chlorine   99–102, 132–3, 147
chromatography   43
coal   24
collision theory   194
competition reaction   112
composite materials   236–7
compound   38
condensation   5, 16
conductor   173
conglomerate   27
contact process   240
cooling curve   4
copper   105, 226–7, 230
core   21–2
coulomb   180
covalent bond   65–8, 138, 204, 208
covalent compound   77
cracking   233
crust   22, 29, 31
crystallisation   42, 239

D

dehydration reaction   156, 215
density   16
diffusion   6, 15
displacement reactions   111–13
distillation   42
  fractional   45–6, 227
Downs cell   176
dyke   23
dynamic equilibrium   198

E

earthquake   21, 31
electrolysis   81–3, 94, 173, 178, 180, 225, 226–7
electrolyte   173, 178